Everyday
Mathematics®

The University of Chicago School Mathematics Project

Student Math Journal
Volume 1

Grade **6**

Wright Group

The University of Chicago School Mathematics Project (UCSMP)

Max Bell, Director, UCSMP Elementary Materials Component; Director, *Everyday Mathematics* First Edition;
James McBride, Director, *Everyday Mathematics* Second Edition; Andy Isaacs, Director, *Everyday Mathematics*
Third Edition; Amy Dillard, Associate Director, *Everyday Mathematics* Third Edition

Authors
Max Bell, John Bretzlauf, Amy Dillard, Robert Hartfield, Andy Isaacs, James McBride, Ann McCarty*,
Kathleen Pitvorec, Peter Saecker, Robert Balfanz†, William Carroll†

Third Edition only   †*First Edition only*

Technical Art
Diana Barrie

Teacher in Residence
Denise Porter

Photo Credits
Cover (l)Stuart Westmoreland/CORBIS, (c)Digital Vision/Getty Images, (r)Kelly Kalhoefer/Getty Images;
Back Cover Digital Vision/Getty Images; iii (t)The McGraw-Hill Companies, (b)CORBIS; iv (t)Microzoa/Getty
Images, (b)Rick Gayle Studio/CORBIS; v Richard T. Nowitz/CORBIS; vi Ryan McVay/Getty Images;
vii The McGraw-Hill Companies; viii CORBIS; 49 59 (l)The McGraw-Hill Companies; 98 Bettmann/CORBIS;
104 FPG/Getty Images.

Contributors
Ann Brown, Sarah Busse, Terry DeJong, Craig Dezell, John Dini, James Flanders, Donna Goffron,
Steve Heckley, Karen Hedberg, Deborah Arron Leslie, Sharon McHugh, Janet M. Meyers, Donna Owen,
William D. Pattison, Marilyn Pavlak, Jane Picken, Kelly Porto, John Sabol, Rose Ann Simpson, Debbi Suhajda,
Laura Sunseri, Jayme Tighe, Andrea Tyrance, Kim Van Haitsma, Mary Wilson, Nancy Wilson, Jackie Winston,
Carl Zmola, Theresa Zmola

This material is based upon work supported by the National Science Foundation under Grant No.
ESI-9252984. Any opinions, findings, conclusions, or recommendations expressed in this material are
those of the authors and do not necessarily reflect the views of the National Science Foundation.

www.WrightGroup.com

Send all inquiries to:
Wright Group/McGraw-Hill
P.O. Box 812960
Chicago, IL 60681

ISBN 0-07-605273-7

17 18 19 20 QDB 14 13 12 11

The McGraw·Hill Companies

Contents

UNIT 1   **Collection, Display, and Interpretation of Data**

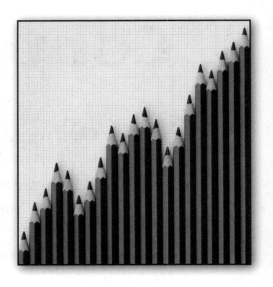

UNIT 2 Operations with Whole Numbers and Decimals

UNIT 3 Variables, Formulas, and Graphs

UNIT 4   Rational Number Uses and Operations

UNIT 5 Geometry: Congruence, Constructions, and Parallel Lines

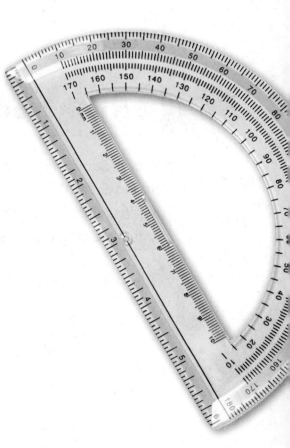

References

Activity Sheets

LESSON 1·1 Sixth Grade Everyday Mathematics

Much of what you learned in the first years of *Everyday Mathematics* served as basic training in mathematics and its uses. In fourth and fifth grades, you built on this training and studied more sophisticated mathematics. This year, you will study new ideas— some of which your parents and older siblings may not have learned until high school. The authors, along with many other people, believe that sixth graders today can learn and do more mathematics than was thought possible 10 or 20 years ago.

Here are a few topics that you will discuss in *Sixth Grade Everyday Mathematics*:

◆ Practice and improve your number sense, measure sense, and estimation skills.

◆ Review and extend your arithmetic, calculator, and thinking skills by working with fractions, decimals, percents, large and small numbers, and negative numbers.

◆ Continue your study of variables, expressions, equations, and other topics in algebra.

◆ Expand your understanding of geometry, with a focus on compass-and-straightedge constructions, transformations of figures, and volumes of 3-dimensional figures.

◆ Explore probability and statistics.

◆ Carry out projects that investigate the uses of mathematics outside the classroom.

A *Student Reference Book* is included in the *Sixth Grade Everyday Mathematics* program. This resource book allows you to look up and review information on topics covered in mathematics both this year and in years past. The *Student Reference Book* also includes the rules of popular mathematical games; a glossary of mathematical terms; and reference information, such as tables of measures, fraction-decimal-percent conversion tables, and place-value charts. (Some of this information also appears at the back of your journal.)

This year's activities will help you appreciate the beauty and usefulness of mathematics. The authors hope you will enjoy *Sixth Grade Everyday Mathematics*. Most importantly, we want you to become more skilled at using mathematics so that you may better understand the world in which you live.

LESSON 1·1 *Student Reference Book* **Scavenger Hunt**

Solve the problems on this page and on the next two pages. Use your *Student Reference Book* to help you.

Write the page number on which you found information in the *Student Reference Book* for each problem. You may not need to look for help in the *Student Reference Book*, but you will earn additional points for showing where you would look if you needed to.

Keep score as follows:

◆ 3 points for each correct answer

◆ 5 points for each correct page number of the *Student Reference Book*

Problem Points	Page Points

1. Draw a diameter of the circle below.   _____ _____

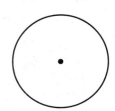

   Student Reference Book, page _____

2. Write the formula for finding the area of a circle.   _____ _____

   Formula: _____

   Student Reference Book, page _____

3. Find the equivalents for the measurements below.   _____ _____

   1 mi = _____ yd

   1 ft^2 = _____ in.^{2}

   1 in. = _____ cm

   1 mi ≈ _____ km

   Student Reference Book, page _____

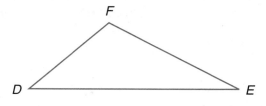

LESSON 1·1   *Student Reference Book* **Scavenger Hunt** *cont.*

	Problem Points	**Page Points**

4.

F

D   E

_____   _____

Name the angles for triangle *DEF.*

_____, _____, _____

Student Reference Book, page _____

5. Write the symbol for *perpendicular.* _____

_____   _____

Student Reference Book, page _____

6. Find the mean for the data set: 24, 26, 21, 18, 26.

_____   _____

mean _____

Student Reference Book, page _____

7. 46 ∗ 53 = _____

_____   _____

Student Reference Book, page _____

8. _____ = 438 + 2,942

_____   _____

Student Reference Book, page _____

9. Is 54,132 divisible by 6? _____

_____   _____

How can you tell without actually dividing?

_____

Student Reference Book, page _____

10. Explain why the number sentence $x + 5 < 8$ is an inequality.

_____   _____

_____

Student Reference Book, page _____

Date _____ Time _____

	Problem Points	Page Points

11. In the decimal number 603,125.748 _____ _____

   a. what is the value of the 8? _____

   b. what digit is in the tenths place? _____

   Student Reference Book, page _____

12. Rename each fraction as a decimal. _____ _____

$$\frac{1}{10} = \rule{2cm}{0.4pt} \qquad \frac{1}{8} = \rule{2cm}{0.4pt}$$

   Student Reference Book, page _____

13. What materials do you need to play *Landmark Shark?* _____ _____

   _____

   _____

   Student Reference Book, page _____

14. What is the height of a geometric solid, such as a prism? _____ _____

   _____

   _____

   Student Reference Book, page _____

15. In Ms. McCarty's class, 9 out of 20 students are boys. _____ _____

   Express the ratio of boys to the total number of students,

   a. using a fraction. _____

   b. using a colon. _____

   Student Reference Book, page _____

Total Problem Points _____	Total Page Points _____

Total Points _____

4

LESSON 1·1   Math Boxes

1. Draw line segments having the following lengths.

    a. $1\frac{1}{4}$ inches

    b. $2\frac{5}{8}$ inches

    c. $\frac{13}{16}$ inch

---

2. Add.

    a. $2{,}653 + 4{,}819 =$ _____
        b. $43{,}708 + 6{,}493 =$ _____

    c. _____ $= 27 + 109 + 75 + 2{,}636$

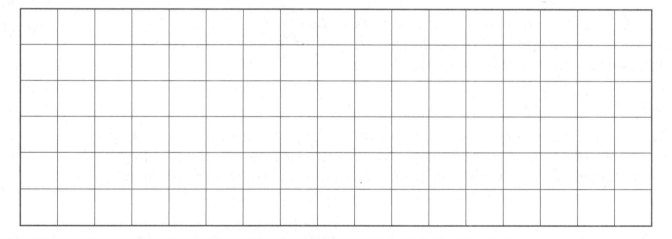

---

3. Use your Geometry Template to draw a polygon whose angles and sides are all the same size.

This type of polygon is called a

_____ .

4. Solve.

    a. $\$4.83 - \$2.96 =$ _____

    b. $\$5.27 + \$6.75 =$ _____

LESSON 1·2 Mystery Plots and Landmarks

Math Message

Complete the following statements. *Do not share your answers.* Estimate if you do not know the exact number.

A. I usually spend about _____ minutes taking a shower or a bath.

B. There is a total of _____ letters in my first, middle, and last names.

C. There are _____ people living in my home.

D. My shoe is about _____ centimeters long (to the nearest centimeter).

E. I watch about _____ hours of television per week.

Mystery Plots

You and your classmates will make **line plots** of the data from the Math Message above. You will then try to figure out which line plot, or **mystery plot,** goes with which statement in the Math Message.

Landmarks

After the class has agreed on the subject of each line plot, mark the number lines for each statement on the next page to show the following data **landmarks** for each statement set: **minimum, maximum, median,** and **mode.** Also record the **range.**

Example:

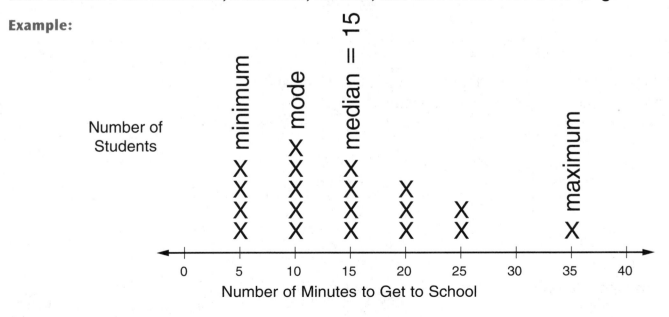

range _____

LESSON 1·2 Mystery Plots and Landmarks *continued*

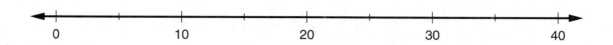

A. Shower/Bath Time (in minutes)   range _____

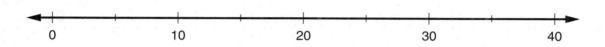

B. Number of Letters in First, Middle, and Last Names   range _____

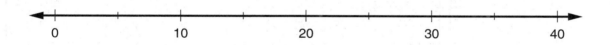

C. Number of People Living in Home   range _____

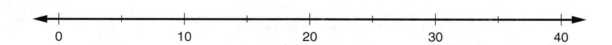

D. Length of Shoe (to nearest cm)   range _____

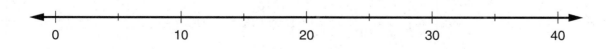

E. Hours of Television Viewed per Week   range _____

LESSON 1·2   **Math Boxes**

---

1. Draw line segments having the following lengths.

   a. 4.2 cm

   b. 51 mm

   c. 42 mm

2. Find the minimum, maximum, median, mode, and range for the set of numbers.

   105, 100, 111, 94, 105

   minimum _____   maximum _____

   median _____   mode _____

   range _____

SRB
136

---

3. Subtract.

   a. 900 − 3 = _____   **b.** 5,182 − 2,637 = _____   **c.** _____ = 8,035 − 675

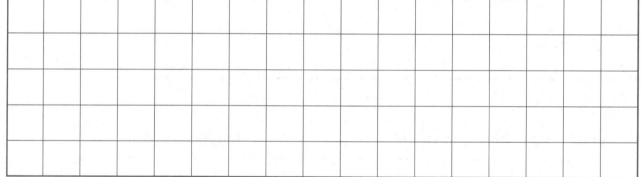

SRB
15–17

---

4. Which is the quotient of 882 and 7?

   Fill in the circle next to the best answer.

   ◯ **A.** 93.6

   ◯ **B.** 126

   ◯ **C.** 12.6

   ◯ **D.** 127

SRB
22–24

5. Complete.

   a. 600 ∗ _____ = 54,000

   b. _____ = 80 ∗ 90

   c. 400 ∗ 80 = _____

   d. 560,000 = 700 ∗ _____

   e. 40 ∗ 700 = _____

SRB
18

LESSON 1·3   # Comparing Graphical Representations

Math Message

a.

Math Test Scores

Stems (100s and 10s)	Leaves (1s)
5	5 6 6 6
6	1 3 3 4 8
7	1 5
8	1 4 4 4 4 7 7 8 8
9	1 5 8 9 9
10	0 0

b.

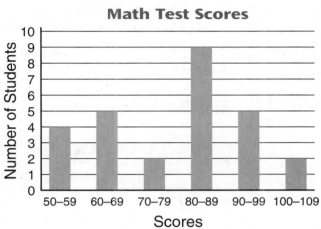

Math Test Scores

c.

Math Test Scores	
Scores	**Number of Students**
50–59	////
60–69	////
70–79	//
80–89	//// ////
90–99	////
100	//

d.

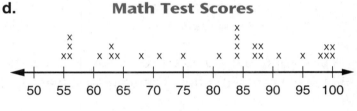

Math Test Scores

1. Use pages 134, 135, and 138 of the *Student Reference Book* to identify each of the data representations (a–d) above.

   a. _____   **b.** _____

   c. _____   **d.** _____

2. Explain how these data representations are alike and how they are different.

   _____

   _____

   _____

3. Which graphical representation helps you identify the range, median, and mode most easily? Explain your choice.

   _____

   _____

LESSON 1·3 Old Faithful Erupts

A geyser is a natural fountain of water and steam that erupts from the ground. Old Faithful is perhaps the most studied geyser of Yellowstone Park. Its eruptions have been recorded since its discovery in 1870. Mathematicians have examined the relationship between the time in minutes an eruption lasts, which is called the duration, and the time to the next eruption, which is called the interval.

Duration data appear in the table below.

Duration of Old Faithful Eruptions (in min)											
(Number of Observations: 48)											
4.9	1.7	2.3	3.5	2.3	3.9	4.3	2.5	3.4	4.8	4.1	1.9
4.6	4.1	2.9	3.7	3.4	1.7	1.7	3.3	4.0	4.6	3.1	2.9
4.1	4.6	2.0	3.5	4.2	4.7	1.8	4.0	1.8	1.9	2.3	2.0
4.5	3.7	3.9	3.9	1.9	4.3	3.2	4.7	3.5	2.0	1.8	4.5

A stem-and-leaf plot is a useful way to find landmarks when there are many data values in random order. The stem-and-leaf plot of the eruption data appears below.

Duration of Old Faithful Eruptions
(Number of Observations: 48)

Stems (ones)	Leaves (tenths)
1	7 7 7 8 8 8 9 9 9
2	0 0 0 3 3 3 5 9 9
3	1 2 3 4 4 5 5 5 7 7 9 9 9
4	0 0 1 1 1 2 3 3 5 5 6 6 6 7 7 8 9

The data in the plot are ordered, making it easier to determine data landmarks. Using the stem-and-leaf plot above, find the minimum, maximum, and range of the duration data.

a. minimum _____   **b.** maximum _____   **c.** range _____

LESSON 1·3 · Stem-and-Leaf Plot: Double Stems

Predicting Old Faithful's eruptions can be difficult. To predict its next eruption, mathematicians have studied the length of time between eruptions, which is called the interval.

Interval data appear in the table below.

Interval of Old Faithful Eruptions (in min)
(Number of Observations: 48)

95	60	49	61	75	68	70	86	58	66	88	93
42	91	45	69	81	57	54	67	80	86	67	83
79	48	50	53	81	77	56	86	72	80	76	53
61	72	88	57	53	51	86	81	77	83	78	70

The stem-and-leaf plot of the interval data has been started for you. Complete the plot by filling in the leaves for each double stem. Remember that for each pair of identical stems, leaves with values of 0–4 go on the upper stem, and leaves with values of 5–9 go on the lower stem.

Interval of Old Faithful Eruptions
(Number of Observations: 48)

Stems (tens)	Leaves (ones)
4	
4	
5	
5	
6	
6	
7	
7	
8	
8	
9	
9	

Use your completed stem-and-leaf plot to find the following landmarks:

a. minimum _____   b. maximum _____   c. range _____

d. mode _____   e. median _____

11

Date _____   Time _____

 LESSON 1·3   # Rounding to Estimate Products

Round each number to its greatest place value.

1. 711; hundreds _____

2. 6,557; thousands _____

3. 22,698; ten-thousands _____

4. 1,943,007; millions _____

5. 34; tens _____

6. 956,391; hundred-thousands _____

Rounding is one strategy you can use to estimate products. One way you can estimate a product is by first rounding each number to its greatest place value and then computing.

Round each number to its greatest place value. Then estimate.

Example: 28 * 52   Rounded factors <u>30</u> * <u>50</u>   Estimate <u>1,500</u>

7. 62 * 79   Rounded factors _____ * _____   Estimate _____

8. 876 * 82   Rounded factors _____ * _____   Estimate _____

9. 456 * 714   Rounded factors _____ * _____   Estimate _____

10. 5,473 * 44   Rounded factors _____ * _____   Estimate _____

11. 3,736 * 633   Rounded factors _____ * _____   Estimate _____

12. 4,892 * 452   Rounded factors _____ * _____   Estimate _____

13. 3,491 * 5,347   Rounded factors _____ * _____   Estimate _____

14. 46,932 * 72   Rounded factors _____ * _____   Estimate _____

15. 16,236 * 284   Rounded factors _____ * _____   Estimate _____

LESSON 1·3 Math Boxes

1. Draw line segments having the following lengths.

   a. $1\frac{3}{4}$ inches

   b. $2\frac{6}{8}$ inches

   c. $\frac{12}{16}$ inch

2. Add.

   a. $4{,}209 + 6{,}385 =$ _____

   b. $472 + 38{,}529 =$ _____

   c. _____ $= 4 + 263 + 1{,}020 + 79$

   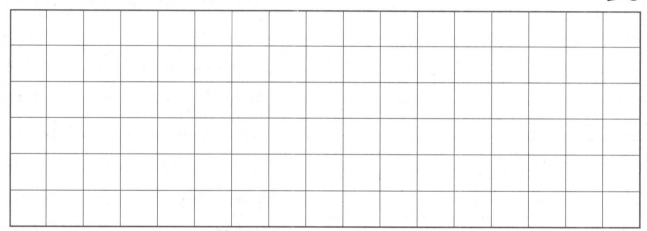

3. Use your Geometry Template to draw a regular hexagon. Then divide this figure into 6 congruent triangles.

   What kind of triangles are these? Circle the best answer.

   A. scalene triangles

   B. equilateral triangles

   C. isosceles triangles

   D. right triangles

4. Solve.

   a. $\$7.22 - \$3.43 =$ _____

   b. $\$9.28 + \$2.76 =$ _____

LESSON 1·4 Comparing the Median and Mean

Math Message

A small office supply company has 12 employees. Their yearly salaries appear in the table, as well as the mean, median, and mode salaries.

1. Does the mean, the median, or the mode best represent the typical salary at the company? Explain.

_____

_____

_____

_____

_____

_____

_____

_____

Salaries at Fancy Font Office Supplies	
Job	**Annual Salary**
President	$275,000
Vice-President: Marketing	$185,000
Vice-President: Sales	$185,000
Marketing Manager	$62,500
Product Manager	$59,000
Sales Manager	$55,500
Promotions Manager	$52,500
Salesperson 1	$45,000
Salesperson 2	$43,500
Salesperson 3	$43,000
Administrative Assistant 1	$39,500
Administrative Assistant 2	$36,000
Mean salary	**$90,125**
Median salary	**$54,000**
Mode salary	**$185,000**

Median and Mean

Find the median and mean for each of the following sets of numbers.

2. 6, 9, 10, 15   a. median _____   b. mean _____

3. 0.50, 0.75, 1, 1.25, 0.80   a. median _____   b. mean _____

4. 123, 56, 92, 90, 88   a. median _____   b. mean _____

LESSON 1·4 Data Landmarks

The 10 most successful coaches in the history of the National Football League (NFL) are listed in the table at the right, along with the number of games won through the end of the 2002 season.

Find the following landmarks for the data set displayed in the table.

1. median _____

2. maximum _____

3. minimum _____

4. mean _____

5. mode _____

6. range _____

Most Successful NFL Coaches	
Coach	**Games Won**
Don Shula	347
George Halas	324
Tom Landry	270
Curly Lambeau	229
Chuck Noll	209
Dan Reeves	198
Chuck Knox	193
Paul Brown	170
Bud Grant	168
Steve Owen	155

SRB
136 137

Try This

7. Denzel's first three test scores in math were 90, 100, and 90.

   a. What must Denzel score on his fourth test to keep his *mean* score at 90 or higher?

   _____

   b. What must Denzel score on his fourth test to keep his *median* test score at 90 or higher?

   _____

   _____

LESSON 1·4   Math Boxes

1. Measure the line segment below to the nearest centimeter.

  a. _____

  _____ cm

Measure the line segment below to the nearest millimeter.

  b. _____

  _____ mm

2. Write a data set that fits the following description.

There are 7 numbers in the data set.
The minimum is 17.
The range is 45.
The median is 32.
The mode is 41.

_____

SRB
136

3. Subtract.

  a. 1,000 − 25 = _____   **b.** 2,037 − 294 = _____   **c.** _____ = 7,214 − 6,218

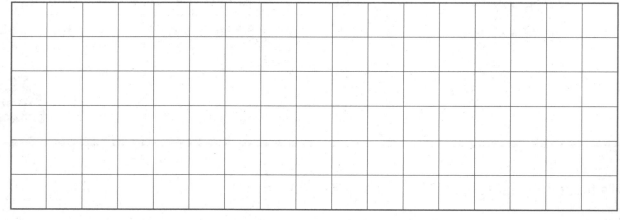

SRB
15–17

4. Find the quotient.

17)459

459 ÷ 17 = _____

SRB
22–24

5. Complete.

  a. 800 * _____ = 48,000

  b. _____ = 60 * 40

  c. 1,500 = 50 * _____

  d. 630,000 = 900 * _____

  e. 90 * 300 = _____

SRB
18

LESSON 1·5 **Math Boxes**

---

1. Hadley's quiz scores were 82, 54, 93, 84, and 92.

   Use the number line below to make a line plot for Hadley's scores.

   Hadley's Quiz Scores

   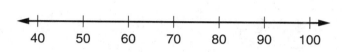

   40   50   60   70   80   90   100

   a. Find the mean score. _____

   b. Does the median or the mean give a better picture of Hadley's overall performance?

      _____

      SRB 134–137

2.

Math Test 1	
Stems (100s and 10s)	**Leaves** (1s)
5	7
6	3 4 8
7	0 0 5 5 5 6 7 8
8	1 4 4 4 4 7 7 8 8
9	2 3 5 9
10	0 0

Math Test 2	
Stems (100s and 10s)	**Leaves** (1s)
6	1 2 2 4 5 6 9 9
7	2 2 3 6 8 9
8	1 1 1 3 6 7
9	0 2 2 4 5 5 6

Which set of scores has the greater median? Circle the best answer.

A. Math Test 1   **B.** Math Test 2   **C.** The medians are equal.

SRB 134–136

---

3. Convert.

   a. 480 cm = _____ m

   b. 0.88 m = _____ cm

   c. 39 cm = _____ m

   d. 1.08 m = _____ cm

   SRB 210

---

4. Complete.

   a. 1,500 / 5 = _____

   b. 8,100 / _____ = 9

   c. _____ / 90 = 50

   d. 28,000 / 70 = _____

   e. _____ = 2,100 / 3

   SRB 21

17

LESSON 1·6 The Climate in Omaha

Omaha, the largest city in Nebraska, is located on the eastern border of the state on the Missouri River.

Precipitation is moisture that falls as rain or snow. Rainfall is usually measured in inches; snowfall is usually translated into an equivalent amount of rain.

**Average Number of Days in Omaha with
At Least 0.01 Inch of Precipitation**

Number of days	Jan	Feb	Mar	Apr	May	Jun	Jul	Aug	Sep	Oct	Nov	Dec
	7	6	7	10	12	11	9	9	9	7	5	7

These averages are the result of collecting data for more than 58 years.

1. Complete the following graph.
   First make a dot for each month to represent the data in the table.
   Then connect the dots with line segments. The result is called a **broken-line graph.**
   This type of graph is often used to show trends.

**Average Number of Days in Omaha with
At Least 0.01 Inch of Precipitation**

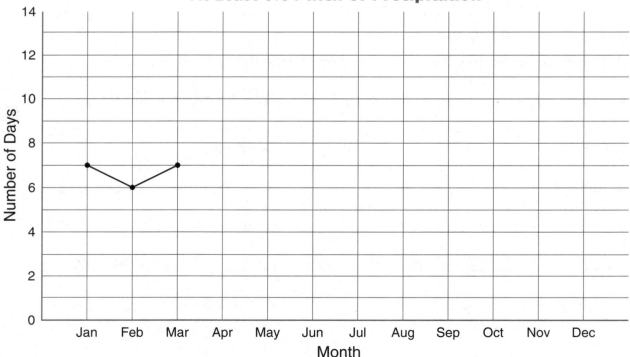

Source: The Times Books World Weather Guide

18

**LESSON
1·6**   # The Climate in Omaha *continued*

Averages of Daily High and Low Temperatures in Omaha, Nebraska

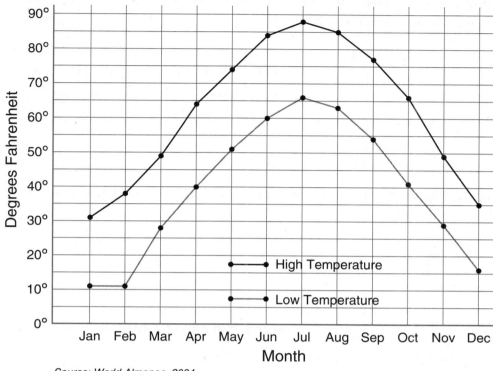

Source: World Almanac, 2004

Here is more information about the climate in Omaha. Black line segments connect the dots for high temperatures. Gray line segments connect low temperatures.

2. On average, what is the

a. warmest month of the year? _____

b. coldest month of the year? _____

3. Compare the average daily high and low temperatures in April.

About how many degrees warmer is the high temperature? _____

4. Use the graph to fill in the missing data in the table below.

Month	Average Daily High Temperature	Month	Average Daily Low Temperature
January		April	
November			41°F
	74°F		60°F
	64°F	March	

LESSON 1·6   Math Boxes

1. The coldest temperature on Earth was recorded at the Russian research station in Vostok, Antarctica. The average temperatures in Vostok for 2002 are shown in the table below.

Month	Jan	Feb	Mar	Apr	May	Jun	Jul	Aug	Sep	Oct	Nov	Dec
Temperature (°F)	−29	−46	−57	−62	−59	−66	−65	−72	−68	−56	−44	−34

Use the data table to complete the broken-line graph below.

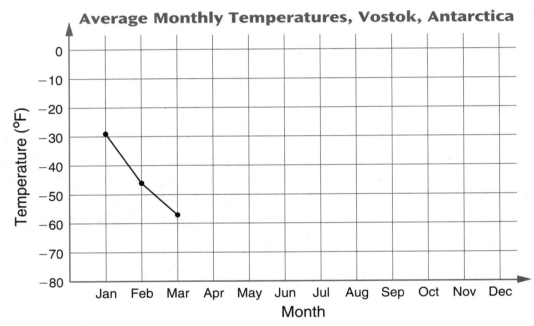

2. Estimate the product 57 * 34.

About _____

Find the exact answer to 57 * 34.

_____ = 57 * 34

3. Use estimation to insert the decimal point in each product.

a. 1.2 * 3 = 3 6

b. 20.2 * 6 = 1 2 1 2

c. 3.8 * 2.6 = 9 8 8

LESSON 1·7 Drawing and Reading Bar Graphs

Math Message

1. Mr. Barr gave his class of 25 students a quiz with 5 questions on it.

   ◆ Every student answered at least 2 questions correctly.

   ◆ Three students answered all 5 questions correctly.

   ◆ Ten students answered 4 questions correctly.

   ◆ The same number of students who answered 2 questions correctly answered 3 questions correctly.

   Draw a bar graph to show all of this information. Title the graph and label each axis.

Answer the following questions about the bar graph to the right.

2. Which ski area had the greatest snowfall in January 1996? _____

3. About how many inches of snow did Keystone receive in January 1996? _____

4. What is the average January snowfall in Loveland? _____

5. How do the average January snowfalls at Loveland and Arapahoe Basin compare?

_____

6. In January 1996, Loveland received how many more inches of snow than Arapahoe Basin? _____

7. Which ski area received the least amount of snow during January 1996? _____

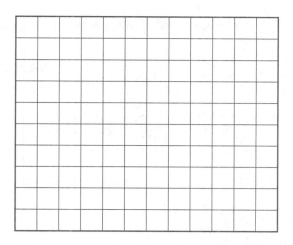

January Snowfalls

Source: Colorado Ski Country, U.S.A

LESSON 1·7

Side-by-Side and Stacked Bar Graphs

SRB
139

Weather in Some Cities in the United States

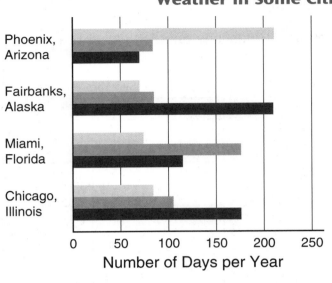

Phoenix, Arizona

Fairbanks, Alaska

Miami, Florida

Chicago, Illinois

0   50   100   150   200   250
Number of Days per Year

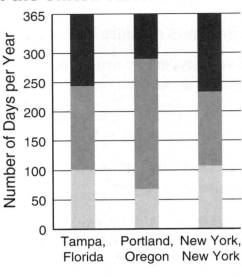

Number of Days per Year

365
300
250
200
150
100
50
0

Tampa, Florida   Portland, Oregon   New York, New York

☐ Clear   ▓ Partly Cloudy   ■ Cloudy

Source: World Almanac 1999

Use the side-by-side graph to answer Problems 1–3. Use the stacked bar graph to answer Problems 4–6. Circle the correct answers.

1. About how many cloudy days does Fairbanks have yearly?

a. 70

b. 85

c. 210

2. Which city ranks second in the number of clear days per year?

a. Chicago

b. Miami

c. Phoenix

3. Which of these pairs of cities have the most similar weather?

a. Phoenix and Fairbanks

b. Fairbanks and Chicago

c. Miami and Chicago

4. About how many clear days does Portland have yearly?

a. 50

b. 70

c. 100

5. Which city has the greatest number of partly cloudy days?

a. Portland

b. Tampa

c. New York

6. About how many cloudy days does New York have yearly?

a. 110

b. 130

c. 230

22

Date _____   Time _____

LESSON 1·7   Temperatures above Earth's Surface

When you fly in a commercial airplane, you probably don't notice any large changes in temperature, because the cabin is temperature controlled. But if you are riding in a hot-air balloon, you feel the temperature drop as the balloon rises.

The table below shows temperatures at various heights above Omaha, Nebraska. The measurements were made on a January day when the temperature at ground level was 32°F. *Reminder:* 1 mile = 5,280 feet.

Height above Ground (miles)	Temperature (°F)
0 (ground level)	32°
1	25°
2	15°
3	−5°
4	−25°
4.5	−40°
5.5	−60°
10	−75°

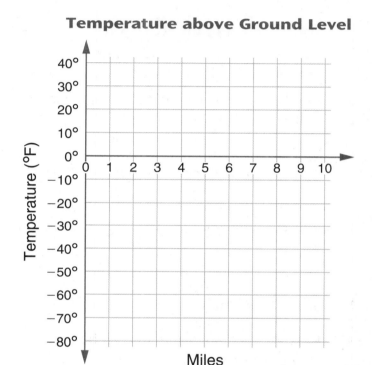

Temperature above Ground Level

1. Make a broken-line graph of the data in the table.

2. At about what height above the ground was the temperature 0°F? _____

3. What was the approximate temperature 8 miles above Omaha? _____

4. Suppose an airplane was about 26,000 feet above Omaha on the day recorded in the table.

   a. What was the approximate temperature at 26,000 feet? _____

   b. How much colder was the temperature at 26,000 feet above ground than at ground level? _____

23

LESSON 1·7   Math Boxes

1. Bigfoot is said to be a large, hairy, humanlike creature that lives in the wilderness areas of the United States and Canada. The following data are some footprint lengths reported by people who claim to have seen Bigfoot in California.

Reported Footprint Lengths (in cm)							
36	46	36	38	33	40	37	56

Use the number line to construct a line plot of the data in the table above.

Footprint Lengths

30   40   50   60

a. Find the mean length. _____

b. Is the median or the mean a better representation of the recorded footprint lengths? Explain.

_____

_____

_____

_____

2. Average wind speeds for major U.S. cities are displayed in the table below.

City	State	Wind Speed (mph)
Anchorage	AK	7.1
Phoenix	AZ	6.2
Denver	CO	8.6
Washington	DC	9.4
Des Moines	IA	10.7
Chicago	IL	10.3
Boston	MA	12.4
Helena	MT	7.7
Memphis	TN	8.8
Salt Lake City	UT	8.8

Construct a stem-and-leaf plot of the wind-speed data.

Average Wind Speeds for U.S. Cities

Stems (10s and 1s)	Leaves (10ths)

SRB 135

3. Convert.

a. 18 cm = _____ mm

b. 120 mm = _____ cm

c. 254 cm = _____ mm

d. 1,020 mm = _____ cm

SRB 210

4. 600,000 / 5,000 = _____

Circle the letter next to the best answer.

A 120,000   **B** 12,000

C 1,200   **D** 120

SRB 21

LESSON 1·8   The Cost of Mailing a Letter

Math Message

The graph below is called a **step graph.** It shows the cost of sending a letter weighing 1 ounce or less by first-class mail anywhere in the United States.

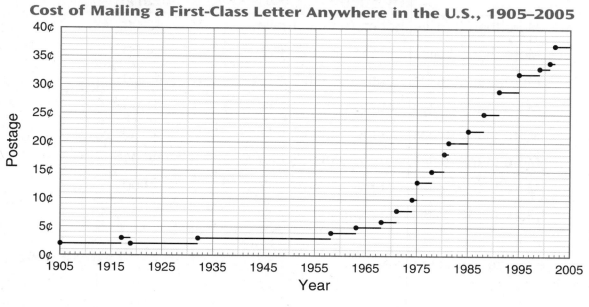

Cost of Mailing a First-Class Letter Anywhere in the U.S., 1905–2005

1. The cost of mailing a letter was 3 cents from 1932 until _____.

2. Did the cost of mailing a letter ever go down? If so, when? _____

3. How many rate increases were there between 1960 and 1970?

   _____

4. In which year did the *greatest* increase occur? _____

5. By how much did the cost of mailing a letter increase between 1960 and 1970? _____

6. Can you tell from the graph the cost of mailing a letter before 1900? _____

7. In which year do you think the cost of mailing a letter will reach 50 cents? Explain.

   _____

   _____

   _____

   _____

LESSON 1·8   Taxicab Fares

The A-1 Taxicab Company charges for a ride according to the distance covered.
Here is the company's table of fares.

Distance	Fare
More than 0 miles, not more than 1 mile	$2.00
More than 1 mile, not more than 2 miles	$4.00
More than 2 miles, not more than 3 miles	$5.00
More than 3 miles, not more than 4 miles	$6.00
More than 4 miles, not more than 5 miles	$7.00
More than 5 miles, not more than 6 miles	$8.00
More than 6 miles, not more than 7 miles	$9.00
More than 7 miles, not more than 8 miles	$10.00

1. Use the data in the table to complete
   the step graph at the right.

2. What is the cost of the fare for each
   distance below?

   a. 3.7 miles   _____

   b. 7.1 miles   _____

   c. 0.1 mile   _____

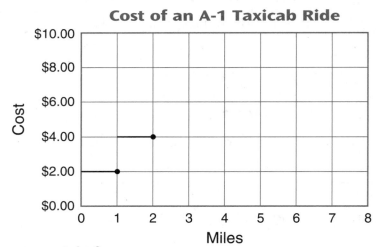

Cost of an A-1 Taxicab Ride

3. You have $7.50. What is the longest trip you can take?

   A. 4 miles   **B.** 5 miles   **C.** 4.9 miles

4. Your ride costs $5.00.

   a. What is the shortest trip you could have taken?

      A. 2 miles   **B.** 2.1 miles   **C.** 3 miles

   b. What is the longest trip you could have taken?

      A. 3 miles   **B.** 2.9 miles   **C.** 3.1 miles

Date _____ Time _____

A Plumber's Rates

The step graph below shows the cost of hiring a plumber from the Drain-Right service for various amounts of time.

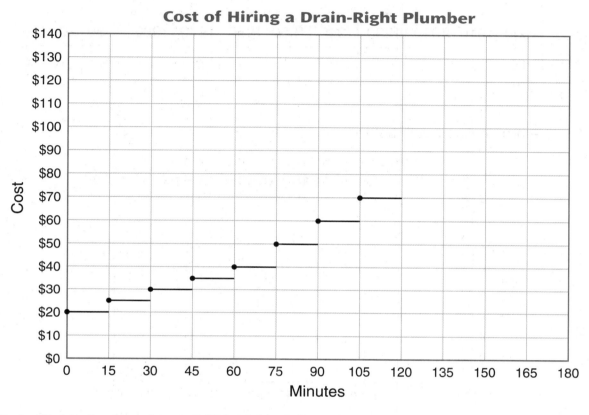

Cost of Hiring a Drain-Right Plumber

1. A Drain-Right plumber charged $80.00 for doing a job.

   a. What is the shortest amount of time the job could have taken?

      A. 121 minutes   **B.** 135 minutes   **C.** 120 minutes

   b. What is the longest amount of time the job could have taken?

      A. 135 minutes   **B.** 134 minutes   **C.** 120 minutes

2. Use the graph to fill in the table below.

Time (min)	20	59	12	120	60	96
Cost						

3. Notice the pattern of the step graph between 60 and 120 minutes. Use this pattern to complete the step graph for all times between 120 and 180 minutes.

27

Date _____ Time _____

1·8   ## Math Boxes

1. The table shows the percent of the U.S. population that was born outside the United States during each decade of the 20th century. Percents have been rounded to the nearest whole.

Year	1900	1910	1920	1930	1940	1950	1960	1970	1980	1990	2000
Percent	14	14	13	12	9	7	5	5	6	8	10

Source: U.S. Census Bureau

Complete the broken-line graph below.

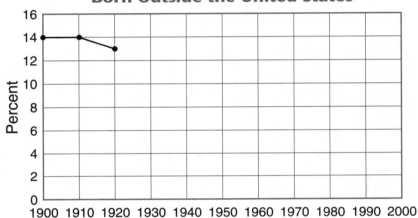

Percent of U.S. Population Born Outside the United States

How would you describe the pattern or trend shown by the graph?

_____

_____

_____

2. Multiply.

a. 81 * 13 = _____

b. _____ = 243 * 72

3. Which of the following estimates is most reasonable for the product 8.3 * 4.7? Fill in the circle next to the best answer.

○ **A.** About 39   ○ **C.** About 13

○ **B.** About 4   ○ **D.** About 3

28

LESSON 1·9 Reading Circle Graphs

Use your estimation skills and your Percent Circle to answer the following questions.

Methods of Preparing Eggs

1. The circle graph at the right shows people's preferences for preparing eggs.

   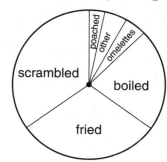

   Source: American Egg Board

   a. Which is the most popular method for preparing eggs? _____

   b. Which is the least popular method for preparing eggs? _____

   c. Which method is preferred by about 25% of the people surveyed? _____

   d. Which methods are preferred by less than 10% of the people surveyed?

   _____

   e. What percent of people surveyed prefer scrambled eggs? About _____%

   f. Is the percent greater for people favoring fried eggs or boiled eggs? _____

      About how much greater? _____%

   g. How do you like your eggs prepared? _____

Pizza Crust Preferences

2. The circle graph at the right shows people's preferences for pizza crust.

   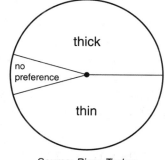

   Source: Pizza Today

   a. True or false: In this survey, the number of people who prefer thin-crust pizza is about the same as the number of people who prefer thick-crust pizza.

   _____

   b. What percent of people prefer thin-crust pizza? About _____%

   c. What percent prefer thick-crust pizza? About _____%

   d. What percent have no preference? About _____%

   e. What kind of crust do you prefer? _____

LESSON 1·9 A Magazine Survey

An issue of a sports magazine for kids featured a readers' survey. In the survey, readers were asked to respond to the following three questions:

1. Should girls be allowed to play on boys' teams?

2. Should boys be allowed to play on girls' teams?

3. On how many organized sports teams do you play during a year?

Readers' responses are represented by the circle graphs below.

Question 1: Should girls be allowed to play on boys' teams?

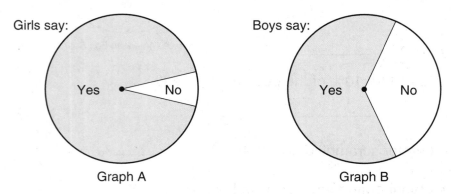

Girls say: Boys say:

Graph A   Graph B

Question 2: Should boys be allowed to play on girls' teams?

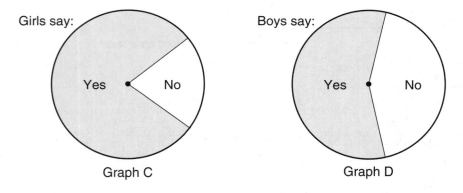

Girls say: Boys say:

Graph C   Graph D

Question 3: On how many organized sports teams do you play during a year?

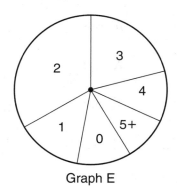

Graph E

LESSON
1·9   **A Magazine Survey** *continued*

Refer to the circle graphs on the preceding page to answer the following questions.

1. Estimate the percent of boys and girls who gave the following responses.

   a. Approximately _____% play on 2 organized teams during a typical year.

   b. Approximately _____% play on *at least* 2 organized teams during a year.

   c. Approximately _____% of the girls think that girls should *not* be allowed
      to play on boys' teams.

2. Which graph (A, B, C, or D) shows almost everyone agreeing on an answer? _____

3. Which graph (A, B, C, or D) shows opinions that are almost evenly divided
   between yes and no?   _____

4. Do you think that the readers of the sports magazine who responded to this survey
   play sports more often, less often, or about the same amount of time as the
   students in your school? Explain.

   _____

   _____

   _____

   _____

   _____

5. Circle graphs on journal page 30 show how boys and girls responded to a survey about
   boys playing on girls' teams and girls playing on boys' teams. Explain how girls' and
   boys' responses to the survey are alike and how they are different.

   _____

   _____

   _____

   _____

LESSON 1·9   Math Boxes

1. Make a step graph from the data in the table.

 SRB 141

Years	1969–1976	1977–1992	1993–1997	1998–2005
Number of Major League Teams	24	26	28	30

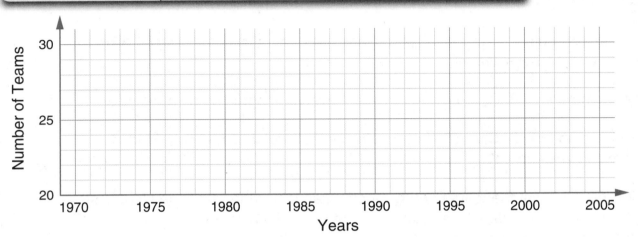

2. According to the circle graph at the right:

   a. Which kind of peanut butter do most Americans prefer? _____

   b. Which kind do about $\frac{1}{3}$ of Americans prefer? _____

   c. About what percent of Americans prefer natural peanut butter? _____

   Peanut Butter Survey

   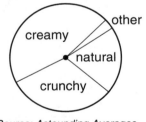

   Source: Astounding Averages

    SRB 145

3.

Tour Number	1	2	3	4	5	6
Number of People	10	9	40	12	14	5

   Does the median or mean best describe the typical size of the tour groups shown in the table above? Explain.

   _____

   _____

   _____

   _____

4. Write the value of the digit 9 in each numeral below.

   a. 401,297 _____

   b. 1,927,387

   _____

   c. 4.95 _____

 SRB 4 28

LESSON 1·10   **Math Boxes**

1. Use the bar graph to answer the questions.

   a. How tall was George Washington?

   _____

   b. How tall was Zachary Taylor?

   _____

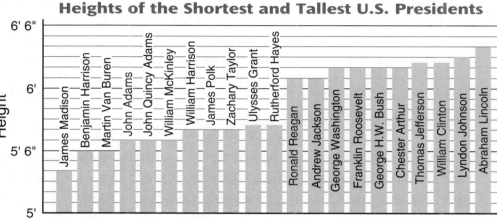

Heights of the Shortest and Tallest U.S. Presidents

Source: The Top 10 of Everything

   c. Who was the shortest president? _____

   d. Who was the tallest president? _____

   e. What is the difference in height between the tallest and shortest president?

   _____

   f. Which height occurs most often for the presidents listed? _____

   g. What is the difference in height between Ulysses Grant and Chester Arthur?

   _____

SRB 138

---

2. Divide. 12)‾3‾0‾0‾

   300 ÷ 12 = _____

SRB 22–24

3. Use estimation to insert the decimal point in each product.

   a. $2.00 × 10 = $2 0 0 0

   b. $0.90 × 10 = $9 0 0

   c. $4.55 × 10 = $4 5 5 0

SRB 35

33

LESSON 1·10 Using a Graph to Find the Largest Area

Math Message

Suppose you have enough material for 22 feet of fence. You want to enclose the largest possible rectangular region with this fence.

1. If you know the perimeter and length of a rectangle, how can you find its width?

   _____

   _____

2. The table below lists the lengths of some rectangles with a perimeter of 22 feet. Fill in the missing widths and areas. You may want to draw the rectangles on your grid paper. Let the side of each grid square represent 1 foot.

Length (ft)	1	2	3	4	5	6	7	8	9	10
Width (ft)	10	9								
Perimeter (ft)	22	22	22	22	22	22	22	22	22	22
Area (ft²)	10	18				28				

3. If you have not already done so, draw each rectangle from the table above on your grid paper.

4. a. What are the length and width of the rectangle(s) in the table with the largest area? _____ ft _____ ft

   b. What is that area? _____ ft²

5. On the graph on page 35, plot the length and area of each rectangle in the table. *Do not connect the plotted points yet.*

6. a. Find a rectangle whose perimeter is 22 feet and whose area is larger than the area of the rectangle in Problem 4 above. Use your graph to help you.

   What is the length? _____ ft Width? _____ ft Area? _____ ft²

   b. Now plot the length and area of this rectangle on the grid on page 35. Then draw a curved line through all the plotted points.

LESSON 1·10 Using a Graph to Find the Largest Area *cont.*

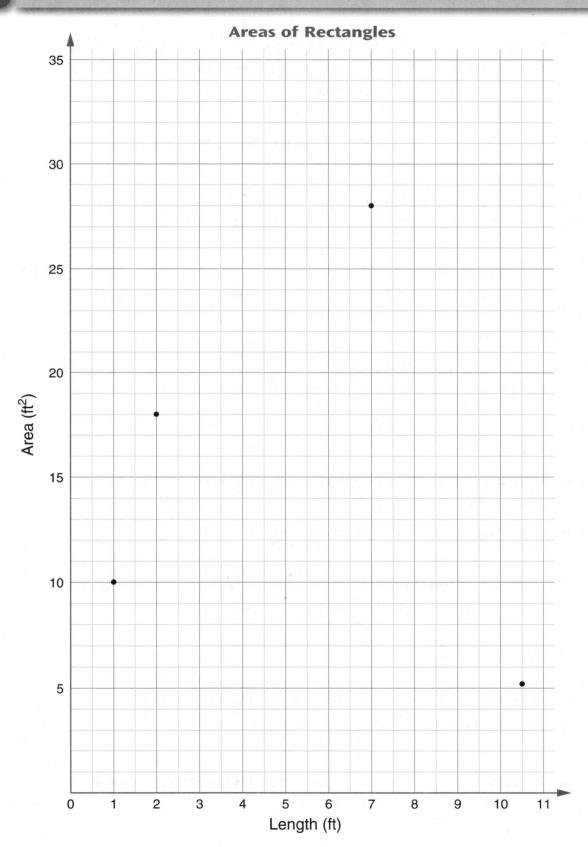

Areas of Rectangles

LESSON 1·11   Statistics Meant to Astound the Reader

Americans Consume 90 Acres of Pizza per Day!
Each day we eat the equivalent of 90 football fields covered with pizza!
The National Association of Pizza Operators reported today that

Ninety acres may seem like a tremendous amount of pizza. The person who wrote this headline wants us to think so. However, let's look at this statistic more closely.

 There are 43,560 square feet in an acre, so 90 acres is about 3,900,000 square feet of pizza.

$90 * 43,560 = 3,920,400$
Round to 3,900,000.

 If Americans eat 3,900,000 square feet of pizza each day for 365 days, that is about 1,420,000,000 square feet of pizza per year.

$3,900,000 * 365 = 1,423,500,000$
Round to 1,420,000,000.

 If 1,420,000,000 square feet of pizza is divided by **about** 270,000,000 people in the United States, then each person, on average, eats about 5 square feet of pizza per year.

$1,420,000,000 / 270,000,000 = 5.\overline{259}$
Round to 5.

Suppose an average pizza is about 1 square foot in area. Then each person in the United States eats approximately 5 pizzas per year.

Here is a new headline based on the information above.

An Average American Eats 5 Pizzas per Year!
The National Association of Pizza Operators reported today that

1. Study the headline below.

An Average American Takes about 50,000 Automobile Trips in a Lifetime!

Write a new headline that gives the same information but will not astound the reader.

_____

_____

Date _____   Time _____

Analyzing Persuasive Graphs

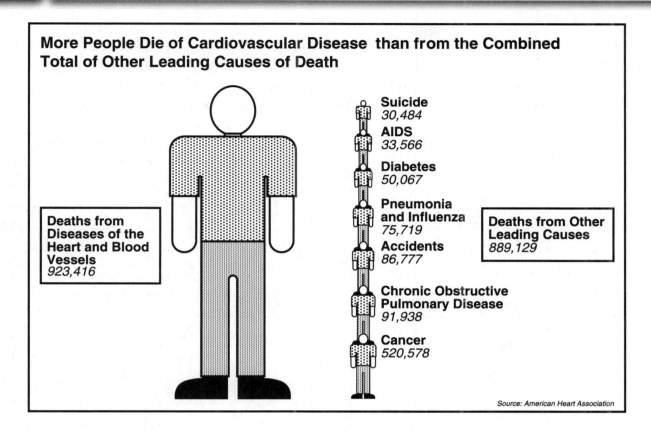

More People Die of Cardiovascular Disease than from the Combined Total of Other Leading Causes of Death

Suicide
30,484

AIDS
33,566

Diabetes
50,067

Pneumonia and Influenza
75,719

Accidents
86,777

Chronic Obstructive Pulmonary Disease
91,938

Cancer
520,578

Deaths from Diseases of the Heart and Blood Vessels
923,416

Deaths from Other Leading Causes
889,129

Source: American Heart Association

1. What is mathematically wrong with the pictograph above?

_____

_____

_____

_____

_____

_____

_____

_____

_____

_____

_____

LESSON 1·11 Analyzing Persuasive Graphs *continued*

You are trying to convince your parents that you deserve an increase in your weekly allowance. You claim that over the past 10 weeks, you have spent more time doing jobs around the house, such as emptying the trash, mowing the lawn, and cleaning up after dinner. You have decided to present this information to your parents in the form of a graph. You have made two versions of the graph and need to decide which one to use.

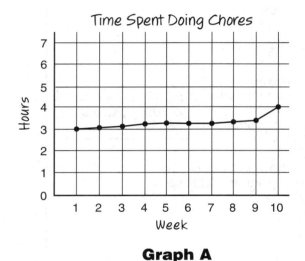

Graph A

Time Spent Doing Chores

Graph B

2. How are Graph A and Graph B similar?

_____

_____

3. How are Graph A and Graph B different?

_____

_____

4. Which graph, A or B, do you think will help you more as you try to convince your parents that you deserve a raise in your allowance? Why?

_____

_____

_____

LESSON 1·11 Math Boxes

1. From 1860 to 1861, 11 states seceded from the Union. All of them were reinstated between 1866 and 1870. Four new states joined the Union between 1860 and 1866. The table at the left shows the number of states in the United States from 1859 to 1870. Make a step graph to display this information.

Years	Number of States
1859	33
1860	32
1861–1862	23
1863	24
1864–1865	25
1866	26
1867	27
1868–1869	34
1870	37

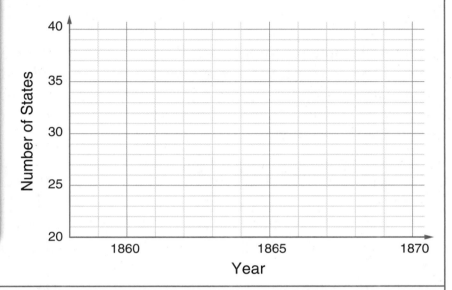

2. Use the circle graphs to answer the following questions.

Percent of Water in Foods

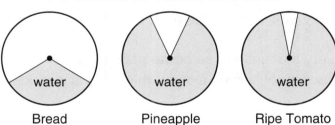

Bread   Pineapple   Ripe Tomato

Source: Astounding Averages

a. About what percent of the content of bread is water?

_____

b. About what percent of the content of a ripe tomato is water?

_____

3. Juan's mean score on three tests was 86. Two of his scores were 92 and 82. What was his third score? Circle the best answer.

A. 76   B. 78

C. 80   D. 84

4. Write the value of the digit 3 in each numeral below.

8.43   _____

24.35   _____

149.073   _____

LESSON 1·12   **Samples and Surveys**

Math Message

A cookie manufacturer that sells chocolate chip cookies uses the sales slogan "65% chips and 35% cookie." Suppose you have a 20-ounce bag of these cookies. Discuss with a partner how you could find what percent of the cookie weight is made up of chocolate chips.

Samples

A **random sample** is a sample that gives all members of the population the same chance of being selected. A **biased sample** is a sample that does not truly represent the total population. A sample is biased when the method used to collect the sample allows some members of the population to have a greater chance of being selected for the sample than others.

Random Sample	Biased Sample
Heights of boys in a sixth-grade class	Heights of boys on the basketball team
Math scores of sixth-grade students	Math scores of students in the math club
Batting average for an entire baseball team	Batting average for starting line-up of a baseball team
Door prize for 1 in every 10 people	Door prize for people who bought 5 or more tickets

For Problems 1–3, tell whether you think the sample is random or biased.

1. Every student whose home phone number ends in 5 was surveyed about plans for the new school library. Do you think this is a random or biased sample? Explain.

   _____

   _____

2. People living in Florida and California were surveyed to determine the percent of Americans who spend at least 1 week of the year at ocean beaches. Do you think this is a random or biased sample? Explain.

   _____

   _____

3. The first 50 people standing in line for a football game were surveyed about how voters feel about a tax increase to build a new stadium. Do you think this is a random or biased sample?

   _____

   _____

40

Date _____ Time _____

Drinking Habits of Children and Teenagers

Read the newspaper article below. Then complete the table and answer the questions.

For years, experts praised milk's nutritional value. However, during the last two decades, many experts have been questioning milk's role in the diet of children who have allergies, sinus problems, or intestinal problems. Based on the results of a comparison study conducted by the USDA (for the years 1978 and 1994), many children are consuming less milk and choosing beverages such as juices and soft drinks instead.

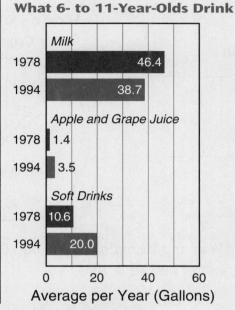

What 6- to 11-Year-Olds Drink

Milk
1978 — 46.4
1994 — 38.7

Apple and Grape Juice
1978 — 1.4
1994 — 3.5

Soft Drinks
1978 — 10.6
1994 — 20.0

Average per Year (Gallons)

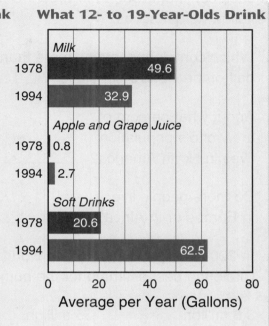

What 12- to 19-Year-Olds Drink

Milk
1978 — 49.6
1994 — 32.9

Apple and Grape Juice
1978 — 0.8
1994 — 2.7

Soft Drinks
1978 — 20.6
1994 — 62.5

Average per Year (Gallons)

Sources: U.S. Dept. of Agriculture; Chicago Tribune

1. Complete the table.

	Ages 6 to 11	Ages 12 to 19
Most popular drink in 1978		
Most popular drink in 1994		
Gallons of soft drinks in 1994		

2. In 1994, 12- to 19-year-olds drank an average of how many gallons of milk per year? _____

3. Were apple and grape juice more popular or less popular in 1994 than they were in 1978?

_____

4. Write *T* for true or *F* for false.

_____ **a.** In 1994, teenagers drank about 3 times as many soft drinks as they did in 1978.

_____ **b.** In 1978 and 1994, teenagers drank more gallons of soft drinks than milk.

_____ **c.** In 1994, both children and teens drank about 1 gallon of milk per week.

LESSON 1·12 Circle Graphs, Median, and Mean

1. The circle graph below shows the approximate percent of the world's population on each continent. Use the Percent Circle on your Geometry Template to answer the following questions.

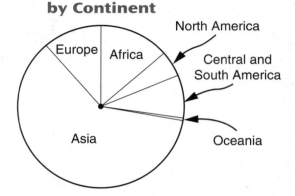

World Population by Continent

a. Which continent contains more than half of the world's population? _____

b. About what percent of the world's population lives in North America? _____

c. Do more people live in Europe or in Africa? _____

d. In 2004, the population of the world was approximately 6,200,000,000 people. Circle the best estimate for the number of people in Asia.

   3.5 million   35 million   350 million   3.5 billion

2. The statements below may or may not be true. Circle the letter of each statement that you can determine is true based on the graph in Problem 1.

a. The population of Europe is more than double the population of North America.

b. In Asia, the population is increasing faster than on any other continent.

c. There are more people in the United States than in Germany.

d. The number of people in Africa is about equal to the total number of people in North, South, and Central America.

3. Elaine tried the long jump 5 times. Her distances were 56 inches, 62 inches, 34 inches, 58 inches, and 62 inches.

a. What is the median of Elaine's long jumps? _____

b. What is the mean? _____

c. Which landmark do you think best represents Elaine's typical jump? _____

LESSON 1·12 Math Boxes

1. Use the bar graph to answer the questions below.

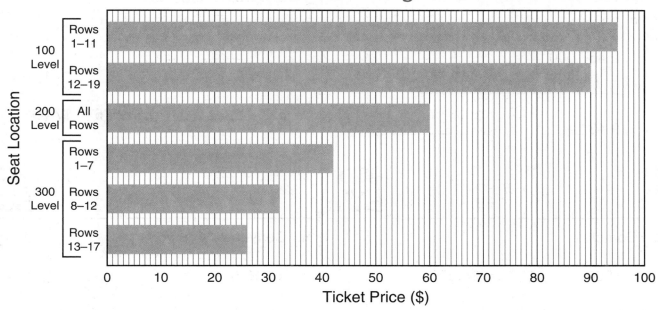

2004–05 Cost of Seats for the Chicago Bulls at the United Center

Seat Location / Ticket Price ($)

- 100 Level — Rows 1–11
- 100 Level — Rows 12–19
- 200 Level — All Rows
- 300 Level — Rows 1–7
- 300 Level — Rows 8–12
- 300 Level — Rows 13–17

a. How much do the cheapest seats at the United Center cost? _____

b. What is the range of prices for seats? _____

c. What is the median price of a 300-level seat? _____

d. Mr. Harris wants to buy four 100-level seats. How much
would he save if he bought the cheaper 100-level
seats rather than the more expensive 100-level seats? _____

SRB
136 138

---

2. Divide. 15)405

405 ÷ 15 = _____

SRB
22–24

3. Which estimate is most reasonable for the
product 10.7 * 4.6?

Choose the best answer.

◯ 0.5   ◯ 50

◯ 5   ◯ 500

SRB
37

LESSON 1·13   Math Boxes

---

1. Write the value of the digit 4 in each numeral below.

a. 551,243 _____

b. 2,457,000

_____

c. 9.48 _____

SRB
4, 28

---

2. Solve.

a. $10.99 + $5.45 = _____

b. _____ = $12.76 − $7.89

SRB
31–33

---

3. Complete.

a. 800 * _____ = 24,000

b. _____ = 40 * 30

c. 500 * 200 = _____

d. _____ = 120,000 * 500

SRB
18

---

4. Convert.

a. 15 m = _____ cm

b. 0.50 m = _____ cm

c. 300 cm = _____ m

d. 25 cm = _____ m

SRB
210

---

5. Use estimation to insert the decimal point in each product.

a. $3.00 * 10 = $3 0 0 0

b. $0.25 * 10 = $2 5 0

c. $8.99 * 10 = $8 9 9 0

SRB
35 36

---

6. Which of the following estimates is the most reasonable for the cost of buying 7 pens that cost $2.25 each?

Choose the best answer.

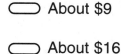 About $9

About $16

About $22

About $25

SRB
37

LESSON 2·1 Reading and Writing Large Numbers

Math Message

A *light-year* is a unit of distance. A light-year is the distance that light travels in one year. One light-year is roughly 9,500,000,000,000 kilometers.

1. Write the number 9,500,000,000,000 in the blank squares of the place-value chart.

trillions			billions			millions			thousands			ones		
100,000,000,000,000	10,000,000,000,000	1,000,000,000,000	100,000,000,000	10,000,000,000	1,000,000,000	100,000,000	10,000,000	1,000,000	100,000	10,000	1,000	100	10	1

Write each of the following numbers in standard notation.

2. one hundred twenty billion _____

3. five trillion, eight hundred seventy-eight billion _____

4. fifty-seven billion, three hundred forty-six million, five hundred three thousand, four hundred nineteen _____

Write the following number in expanded notation.

Example: 9,763 = (9 * 1,000) + (7 * 100) + (6 * 10) + (3 * 1)

5. 47,062

_____

_____

6. 68,250,000

_____

_____

LESSON 2·1 — Number-and-Word Notation

```
H  T  O ' H  T  O ' H  T  O ' H  T  O ' H  T  O
       trillion    billion    million    thousand
```

Convert the numbers given in standard notation to number-and-word notation.

Example: SeaWorld Adventure Park© in Orlando, Florida, is the second most visited aquarium in the United States. Each year, approximately 5,100,000 people visit this attraction. __*5.1 million*__ people

1. Disneyland Paris©, the fourth most popular amusement park in the world, has about 12,800,000 visitors annually. _____ visitors

2. Approximately 8,700,000,000 one-dollar bills are in circulation in the United States. _____ dollar bills

3. Ross, the tenth closest star to Earth, is approximately 61,100,000,000,000 miles away. _____ miles

Convert the numbers given in number-and-word notation to standard notation.

Example: The Andromeda galaxy is 2.3 million light-years away. __*2,300,000*__ light-years

4. The Library of Congress in Washington, D.C., contains more than 128 million items. _____ items

5. Bill Gates purchased one of Leonardo da Vinci's notebooks for 30.8 million dollars. _____ dollars

6. The tuition, books, and living expenses for 4 years of college can be as much as a quarter of 1 million dollars. _____ dollars

LESSON 2·1   **Math Boxes**

1. Circle the value of the underlined digit in the number 1<u>9</u>3,247,056,825.

   A. 900,000,000,000

   B. 90,000,000,000

   C. 9,000,000,000

   D. 900,000,000

   SRB 4

2. Convert the numbers given in standard notation to number-and-word notation.

   a. The Milky Way galaxy is about 150,000 light-years across.

   _____ light-years

   b. The greatest distance that Neptune is from Earth is about 2,920,000,000 miles.

   _____ miles

3. Use the stem-and-leaf plot to find the following landmarks.

   a. maximum _____

   b. minimum _____

   c. median _____

   d. mode _____

Stems (100s and 10s)	Leaves (1s)
20	0 0 5
21	0 4 5 8
22	0 0 0 4 6
23	5 8 9
24	
25	4 6 8 8

   SRB 135 136

4. Complete the "What's My Rule?" table.

   Rule: Multiply by 10.

in	out
8	
27	
403	
3,925	
45,021	

   SRB 253

5. Solve.

   Marta's mother is 5 times as old as Marta. Marta's mother is 25 years old.

   a. How old is Marta?

   _____

   b. In 5 years, Marta's mother will be 3 times as old as Marta. How old will Marta be then?

   SRB 240

LESSON
2·2
Reading and Writing Numbers between 0 and 1

SRB
28

Math Message

A grain of salt is about 0.1016 millimeter long.

Write the number 0.1016 in the place-value chart below.

hundreds	tens	ones	and	tenths	hundredths	thousandths	ten-thousandths	hundred-thousandths	millionths
100	10	1	.	0.1	0.01	0.001	0.0001	0.00001	0.000001
			.						

Write each of the following numbers in standard form.

1. four tenths _____

2. twenty-three hundredths _____

3. seventy-five thousandths _____

4. one hundred nine ten-thousandths _____

5. eight hundredths _____

6. one and fifty-four hundredths _____

7. twenty-four and fifty-six thousandths _____

8. Write the word name for the following decimal numbers.

   a. 0.00016 _____

   b. 0.000001 _____

Date _____   Time _____

Reading and Writing Small Numbers

Complete the following sentences.

Example: A grain of salt is about 0.004, or four _thousandths_, of an inch long.

1. A penny weighs about 0.1, or

   one _____, of an ounce.

2. A dollar bill weighs about 0.035, or

   thirty-five _____, of an ounce.

3. On average, fingernails grow at a rate of about 0.0028, or

   _____ ten-thousandths, of a centimeter per day.

4. Toenails, on average, grow at a rate of about 0.0007, or

   seven _____, of a centimeter per day.

5. It takes about 0.005, or _____,
   of a second for a smell to transfer from the nose to the brain.

6. A baseball thrown by a major-league pitcher takes about 0.01, or

   one _____, of a second to cross home plate.

7. A flea weighs about 0.00017, or _____
   hundred-thousandths, of an ounce.

8. A snowflake weighs about 0.00000004, or

   four _____, of an ounce.

Try This

9. About how many times heavier is a penny than a dollar bill?

   _____

10. About how many times faster do fingernails grow than toenails?

    _____

LESSON 2·2 **Expanded Notation for Small Numbers**

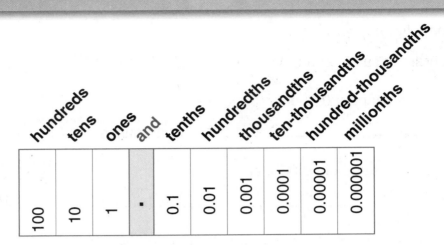

Write each of the following numbers in expanded notation.

Example: 5.96 $(5 * 1) + (9 * 0.1) + (6 * 0.01)$

1. 0.847 _____

2. 3.093 _____

3. 25.3 _____

Give the value of the underlined digit in each number below.

Example: 2.3504 0.05, or 5 hundredths

4. 196.9665 _____

5. 15.9994 _____

6. 23.62173 _____

7. 387.29046 _____

Write each number as a fraction or mixed number.

Example 1: 0.735 $\dfrac{735}{1000}$ **Example 2:** 3.41 $3\dfrac{41}{100}$

8. 17.03 _____   9. 235.075 _____   10. 0.0543 _____

11. Use extended facts to complete the following.

   a. 1 tenth = 1 ÷ _____   **b.** 1 hundredth = 0.1 ÷ _____   **c.** 1 thousandth = 0.01 ÷ _____

50

LESSON 2·2   **Math Boxes**

---

1. Write each of the following numbers using digits.

   a. five and fifty-five hundredths

   _____

   b. one hundred eight thousandths

   _____

   c. two hundred six and nineteen ten-thousandths

   _____

2. Write each number in expanded form.

   a. 53.078 _____

   _____

   b. 9.0402 _____

   _____

---

3. This line graph shows the average monthly rainfall in Jacksonville, Florida.

   Which conclusion can you draw from the graph? Fill in the circle next to the best answer.

   Ⓐ At least 10 months of the year, the average rainfall is less than 3.5 inches.

   Ⓑ The average rainfall increases from June through December.

   Ⓒ The average rainfall for May and November is about the same.

   Ⓓ Jacksonville gets more rain on average than Tampa.

   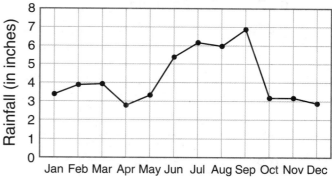

   Average Monthly Rainfall in Jacksonville, Florida

---

4. Janessa is 3 years older than her brother Lamont.

   a. If Janessa is 18 years old, how old is Lamont?

   _____

   b. How old is Janessa when she is twice as old as Lamont?

   _____

5. Find the perimeter of the square if $s = 4.3$ cm. Use the formula $P = 4 * s$, where s represents the length of one side.

   $P =$ _____ cm

LESSON 2·3 Adding and Subtracting Decimals

Estimate each sum or difference. Then solve.

1. 32.5 + 19.6   Estimate _____

32.5 + 19.6 = _____

2. 5.67 − 1.84   Estimate _____

5.67 − 1.84 = _____

3. 10.89 − 3.5   Estimate _____

10.89 − 3.5 = _____

4. 4.07 + 9.38   Estimate _____

4.07 + 9.38 = _____

5.   0.671   Estimate _____
     + 8.935

0.671 + 8.935 = _____

6.   115.97   Estimate _____
       257.49
     + 19.95

115.97 + 257.49 + 19.95 = _____

7.   49.2   Estimate _____
     − 27.6

49.2 − 27.6 = _____

8.   5.006   Estimate _____
     − 0.392

5.006 − 0.392 = _____

9.   8.03   Estimate _____
     − 4.715

8.03 − 4.715 = _____

10.   13.9   Estimate _____
      − 0.38

13.9 − 0.38 = _____

LESSON 2·3   Adding and Subtracting Decimals *continued*

For Problems 12 and 13, round the more precise measurement to match the less precise measurement.

11. In 1999, Maurice Greene set a new world record for the 100-meter dash, running the race in 9.79 seconds. Until then, the record had been held by Donovan Bailey, who ran the race in 9.84 seconds.

    How much faster was Greene's time?

    _____
    (unit)

12. In the 2000 Olympic Games, the United States women's 400-meter medley relay team won this swimming event with a time of 3 minutes, 58.3 seconds. In 2004, the United States women's team took second with a time of 3 minutes, 59.12 seconds.

    How much faster was the 2000 team than the 2004 team?

    _____
    (unit)

13. In the 2000 Olympic Games, Kamila Skolimowska of Poland won the women's hammer throw with a distance of 71.2 meters. In 2004, Olga Kuzenkova of Russia won with a distance of 75.02 meters.

    a. Which woman threw the greater distance? _____

    b. By how much? _____
    (unit)

14. Find the perimeter of the rectangle in meters or centimeters.

    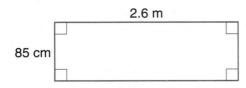

    Perimeter = _____
    (unit)

LESSON
2·3
Math Boxes

1. Write each number in expanded form.

a. 346.52

_____

_____

b. 70.039

_____

_____

2. Convert the numbers given in number-and-word notation to standard notation.

a. The least distance that Pluto is from Earth is about 2.7 billion miles.

_____ miles

b. Earth is about 150 million kilometers from the Sun.

_____ kilometers

3. Make a double-stem plot for the data below.

Inauguration Age of U.S. Presidents Since 1861						
52	56	46	54	49	50	47
55	55	54	42	51	56	55
51	54	51	60	62	43	55
56	61	52	69	62	46	54

Use your double-stem plot to find the following landmarks.

a. range _____   b. median _____   c. mode(s) _____

SRB 135 136

4. Complete the "What's My Rule?" table.

Rule: Multiply by 10.

in	out
7	
80	
0.4	
9.2	
32.6	

SRB 253

5. Solve.

Alma is $\frac{1}{4}$ the age of her father. Alma's father is 60 years old.

a. How old is Alma?

_____

b. Alma's brother Luke is $\frac{1}{5}$ his father's age. How old is Luke?

_____

SRB 240

LESSON 2·3 Drawing and Interpreting a Step Graph

The cost of a telephone call usually depends on how long the call lasts. The calling costs for one phone company appear in the table at the right.

Length of Call	Cost
More than **0** minutes, but not more than **20** minutes	$5.00
More than **20** minutes, but not more than **21** minutes	$5.10
More than **21** minutes, but not more than **22** minutes	$5.20
More than **22** minutes, but not more than **23** minutes	$5.30
More than **23** minutes, but not more than **24** minutes	$5.40

1. Use the data in the table to complete the step graph below.

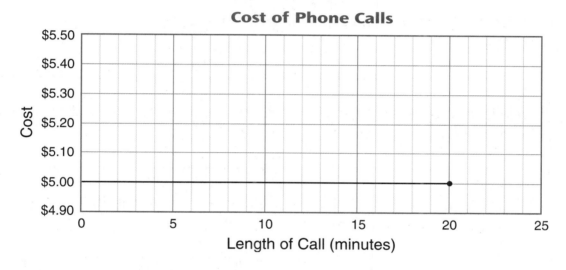

Cost of Phone Calls

Give the cost for each phone call.

2. 1 minute costs

$_____.

3. 12 minutes cost

$_____.

4. 20 minutes cost

$_____.

5. A phone call costs $5.40. Choose the best answer for each of the following questions.

   a. What is the shortest amount of time the phone call can last?

      ⬭ 23 minutes   ⬭ between 23 and 24 minutes   ⬭ 24 minutes

   b. What is the longest amount of time the phone call can last?

      ⬭ 23 minutes   ⬭ between 23 and 24 minutes   ⬭ 24 minutes

6. If you make a 4-minute call, what is the cost per minute? $_____

LESSON 2·4   Multiplying by Powers of 10

Math Message

Some Powers of 10									
10^4	10^3	10^2	10^1	10^0	.	10^{-1}	10^{-2}	10^{-3}	10^{-4}
$10*10*10*10$	$10*10*10$	$10*10$	10	1	.	$\frac{1}{10}$	$\frac{1}{10}*\frac{1}{10}$	$\frac{1}{10}*\frac{1}{10}*\frac{1}{10}$	$\frac{1}{10}*\frac{1}{10}*\frac{1}{10}*\frac{1}{10}$
10,000	1,000	100	10	1	.	0.1	0.01	0.001	0.0001

Work with a partner. Study the table above and then discuss the following questions.

1. What is the relationship between the exponent (Row 1) and the number of zeros in the positive powers of 10 (Row 3)?

   _____

2. What is the relationship between the exponent (Row 1) and the number of digits after the decimal point in the negative powers of 10 (Row 3)?

   _____

3. In Lesson 2-1, you learned that each place on the place-value chart is 10 times the value of the place to its right and $\frac{1}{10}$ the value of the place to its left.

   a. What happens to the decimal point when you multiply 1.0 by 10; 10.0 by 10; 100.0 by 10; and so on?

   _____

   b. What happens to the decimal point when you multiply 1.0 by $\frac{1}{10}$; 10.0 by $\frac{1}{10}$; 100.0 by $\frac{1}{10}$; and so on?

   _____

Use extended multiplication facts and any strategies you have developed through your study and discussion of patterns to complete the following.

4. $0.2 * 10^3 =$ _____

5. $70 = 0.07 *$ _____

6. $0.5 = 0.005 *$ _____

LESSON 2·4

Multiplying by Powers of 10 *continued*

Some Powers of 10								

10^4	10^3	10^2	10^1	10^0	.	10^{-1}	10^{-2}	10^{-3}	10^{-4}
$10*10*10*10$	$10*10*10$	$10*10$	10	1	.	$\frac{1}{10}$	$\frac{1}{10}*\frac{1}{10}$	$\frac{1}{10}*\frac{1}{10}*\frac{1}{10}$	$\frac{1}{10}*\frac{1}{10}*\frac{1}{10}*\frac{1}{10}$
$10{,}000$	$1{,}000$	100	10	1	.	0.1	0.01	0.001	0.0001

Use extended multiplication facts and any strategies you have developed through your
study and discussion of patterns to complete the following.

7. $50 * 0.1 =$ _____

8. $5 * 10^{-1} =$ _____

9. $0.4 * 10^{-1} =$ _____

10. $0.4 * 10^{-2} =$ _____

11. $0.4 * 10^{-3} =$ _____

12. $3.2 * 0.001 =$ _____

13. $0.32 = 3.2 *$ _____

14. $0.032 =$ _____ $* 10^{-2}$

15. In Expanded notation, $5.26 = (5 * 1.0) + (2 * 0.1) + (6 * 0.01)$.
Exponential notation can be used to write numbers in expanded notation.
Using exponential notation, $5.26 = (5 * 10^0) + (2 * 10^{-1}) + (6 * 10^{-2})$.

Write each number in expanded notation using exponential notation.

a. $0.384 =$ _____

b. $71.0295 =$ _____

_____

Recall that percents are parts of 100. You can think of the percent symbol, %, as meaning
times $\frac{1}{100}$, times 0.01, or *times 10^{-2}.*

Example: $25\% = 25 * 10^{-2} = 0.25$

Use your power-of-10 strategy to rewrite each percent as an equivalent decimal number.

16. $5\% =$ _____

17. $50\% =$ _____

18. $500\% =$ _____

19. $125\% =$ _____

20. $12.5\% =$ _____

21. $1\frac{1}{4}\% =$ _____

22. $3\% =$ _____

23. $0.3\% =$ _____

24. $0.03\% =$ _____

Date _____ Time _____

---

1. Write each of the following numbers using digits.

   a. seventy-six hundredths _____

   b. eighty-one and thirty-seven thousandths

   _____

   c. four hundred fifty-three ten-thousandths

   _____

SRB
26 27

2. Write each number in standard form.

   a. $(4 * 1) + (9 * 0.1) + (7 * 0.01)$

   _____

   b. $(6 * 0.1) + (3 * 0.01) + (7 * 0.0001)$

   _____

---

3. Ms. Barrie's science class placed a small amount of sand in direct sunlight for 10 minutes. Every 2 minutes, the students recorded the temperature of the sand. They moved the sand out of the sunlight and continued to record the temperature.

Make a line graph of their data, which is shown in the table at the right. Label each axis of your graph.

Time (minutes)	Temperature (°C)
0	20
2	24
4	28
6	32
8	34
10	35
12	34
14	32
16	30
18	29
20	27

Sand Temperature

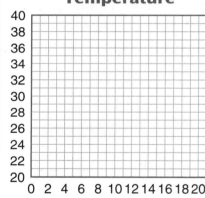

_____

---

4. In the equation $d - 452.23 = 109.76$, the variable d represents the amount of money that was in Zach's savings account at the beginning of the month.

How much money was in Zach's savings account at the beginning of the month?

_____

SRB
250 251

5. The volume of a cube can be found by using the formula $V = s^3$, where s represents the length of one side.

Find the volume of the cube if $s = 3$ cm.

$V = $ _____ cm³

SRB
6 221

LESSON 2·5 Multiplying Decimals

Math Message

1. Diego has 8 quarters. Each quarter is 1.75 mm thick. If Diego stacks the quarters, how high will the stack be?

   a. Express your answer in millimeters. _____ mm

   b. Express your answer in centimeters. _____ cm

The U.S. Mint is responsible for designing and producing our nation's coins. Silver half-dollars have been minted in large quantities since 1793. They became very popular with the introduction of the Kennedy half-dollar in 1964.

2. The Kennedy half-dollar is 2.15 mm thick.

   a. Estimate the height of a stack of 20 half-dollars. Estimate _____ mm

   b. Now find the actual height of the stack in mm. _____ mm

   c. Convert the height from 2b to cm. _____ cm

3. The standard weight of a Kennedy half-dollar is 11.34 g.

   a. Estimate the weight of 8 half-dollars. Estimate _____ g

   b. Calculate the weight of 8 half-dollars. _____ g

4. A U.S. dollar bill is about 6.6 cm wide and 15.6 cm long.

   a. The length of the bill is about _____ times as great as the width.

   b. Estimate the area of a dollar bill. Estimate _____ cm^2

   c. Calculate the area of a dollar bill. _____ cm^2

   d. Explain how you knew where to place the decimal point in your answer to 4c.

   _____

   _____

LESSON 2·5 Multiplying Decimals *continued*

The multiplication for Problems 5–8 has been done for you, but the decimal point has not been placed in the product. Place the decimal point correctly.

5. $2.3 * 7.3 = 1\ 6\ 7\ 9$

6. $51 * 3.8 = 1\ 9\ 3\ 8$

7. $6.91 * 8.2 = 5\ 6\ 6\ 6\ 2$

8. $0.2 * 5.777 = 1\ 1\ 5\ 5\ 4$

9. Explain how you decided where to place the decimal point in Problem 8.

_____

_____

10. Masumi used her calculator to multiply $9.1 * 2.3$. She got the answer 209.3.

a. Explain why this is not a reasonable answer.

_____

_____

b. What do you think Masumi might have done wrong?

_____

_____

Without using a calculator, estimate, and then find each of the following products. Use your estimation skills to help you place each decimal point. Show your work.

11. 2.7   Estimate _____
    * 4.5
    ‾‾‾‾‾

2.7 * 4.5 = _____

12.   24   Estimate _____
    * 5.1
    ‾‾‾‾‾

24 * 5.1 = _____

13. 5.4   Estimate _____
    * 0.2
    ‾‾‾‾‾

5.4 * 0.2 = _____

14. 9.8   Estimate _____
    * 16
    ‾‾‾‾‾

9.8 * 16 = _____

LESSON 2·5   **Math Boxes**

---

1. Multiply mentally.

   a. $0.8 * 10 =$ _____

   b. $77 * 0.1 =$ _____

   c. $0.17 * 1{,}000 =$ _____

   d. $985 * 0.001 =$ _____

SRB 35 36

2. Estimate the product $5.1 * 2.6$.

Estimate _____

Multiply $5.1 * 2.6$. Show your work.

$5.1 * 2.6 =$ _____

SRB 37

---

3. Each day for 5 weeks, Ms. Porter's class recorded the gasoline prices of local gas stations. The class then calculated weekly mean prices, which appear in the table at the right.

Which of the following graphs would be best for showing how the mean price changed over time? Circle the best answer.

   A. Line plot

   B. Circle graph

   C. Broken-line graph

   D. Stem-and-leaf plot

Mean Price of Gasoline

Week	Mean Price (per gallon)
1	$1.85
2	$1.91
3	$2.04
4	$1.99
5	$1.93

---

4. Complete the table for the given rule.

Rule: Subtract the *in* number from 8.3.

in	out
5	
4.1	
	4.5
1.9	

SRB 33 253

5. Plot each number on the number line and write the letter label for the point.

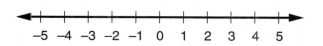

   A: -1 **B:** 3

   C: -2 **D:** 4

SRB 99

LESSON 2·6   **More Multiplying Decimals**

Math Message

Estimate each product. Then use the lattice method to solve the problem.

1. 28 * 13   Estimate _____

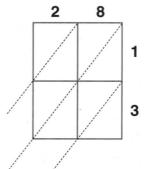

   28 * 13 = _____

2. 2.8 * 1.3   Estimate _____

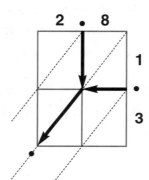

   2.8 * 1.3 = _____

Use the product of 456 and 78 to solve the following problems.

3. 45.6 * 7.8   = _____

4. 456 * 0.78   = _____

5. 4.56 * 78   = _____

6. 0.456 * 7.8   = _____

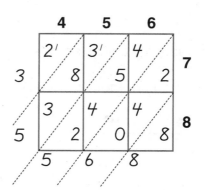

LESSON 2·6

More Multiplying Decimals *continued*

Use any multiplication algorithm you choose to solve the following problems.
Show your work.

7.   3.4
       * 7.9
       _____

8.   0.46
       * 0.83
       _____

9.   19.6
       *   4
       _____

10.   23.65
        *   8
        _____

11.   0.48
        * 25.4
        _____

12.   0.21
        *   26
        _____

13.   4.8
        * 15
        _____

14.   1.52
        * 0.3
        _____

Solve the following problems mentally.

15. 1.5 * 10 = _____

16. 0.25 * 6 = _____

17. _____ = 1.2 * 0.5

18. 0.03 * 6.2 = _____

19. 5.1 * 0.4 = _____

20. _____ = 0.07 * 0.6

21. _____ = 3.42 * 0.2

22. _____ = 4 * 9.1

63

LESSON 2·6 Math Boxes

1. Estimate each sum or difference. Then solve.

   a. $35 - 18.23$ Estimate _____

         $35 - 18.23 =$ _____

   b. $674.92 + 0.843$ Estimate _____

         $674.92 + 0.843 =$ _____

2. $0.53 * 29.6$ Estimate _____

Use any multiplication algorithm to solve $0.53 * 29.6$. Show your work.

$0.53 * 29.6 =$ _____

3. Multiply mentally.

   a. $0.93 * 10^{-1} =$ _____

   b. $4.73 * 10^{2} =$ _____

   c. $0.05 * 10^{-2} =$ _____

   d. $0.08 * 10^{4} =$ _____

4. Which landmark do you think is generally used in each situation?

   a. A teacher reports class scores on a math test.

      Landmark _____

   b. The owner of a small grocery store decides which brand of cereal to stock.

      Landmark _____

5. Plot and label the following points on the coordinate grid.

   A: $(4,5)$ **B:** $(-3,6)$

   C: $(2,-2)$ **D:** $(0,4)$

   E: $(-6,-3)$ **F:** $(1,0)$

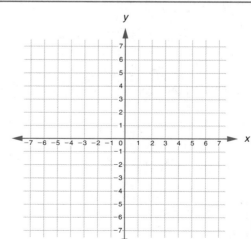

LESSON 2·7   **Math Boxes**

---

1. Multiply mentally.

  a. $0.075 * 10^2 = $ _____

  b. $2,847.6 * 10^{-3} = $ _____

  c. $0.0359 * 10^5 = $ _____

  d. $919 * 10^{-4} = $ _____

35 36

---

2. Estimate the product $14.7 * 0.65$.

  Estimate _____

  Multiply $14.7 * 0.65$. Show your work.

37

  $14.7 * 0.65 = $ _____

---

3. The circle graph shows how Tyler spent the $152 he earned last week. What percent of his earnings did he use for transportation?

  _____

Tyler's Expenses

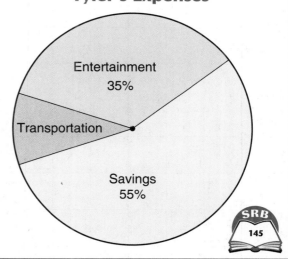

Entertainment 35%

Transportation

Savings 55%

145

---

4. Write the rule for the table in words.

in	out
$47.99	$479.90
$12.10	$121
$0.59	$5.90
$0.08	$0.80

  Rule: _____

35 253

---

5. Match each description with the point on the number line that represents it.

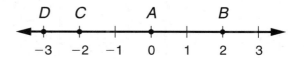

  a. The opposite of 2   _____

  b. The sum of any number and its opposite   _____

99 100

LESSON 2·7 Practicing Division

3 Ways to Write a Division Problem

$246 \div 12 \rightarrow 20$ R6 $12\overline{)246} \rightarrow 20$ R6 $246 / 12 \rightarrow 20$ R6

2 Ways to Express a Remainder

$12\overline{)246} \rightarrow 20$ R6 $12\overline{)246} = 20\frac{6}{12}$, or $20\frac{1}{2}$

Divide.

1. 752 / 23 _____

2. 839 ÷ 58 _____

3. 2,436 ÷ 28 _____

4. 150$\overline{)1,350}$ _____

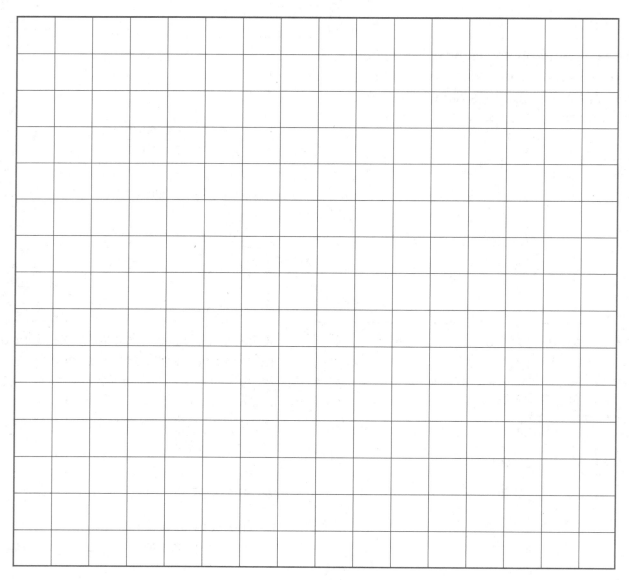

LESSON 2·7 **Practicing Division** *continued*

Solve the following problems mentally or use a division algorithm.

5. The Petronas Twin Towers is an 88-story building in Malaysia. It costs about $60,000 to rent 2,500 square feet of office space in the towers for 1 year. What is the cost per month? About _____

(unit)

6. A professional hockey stick costs about $60. Lucero's team has $546 to use for equipment. How many sticks can the team buy? _____

(unit)

7. In 1650, it took about 50 days to sail from London, England, to Boston, Massachusetts, which is a distance of about 3,700 miles. On average, about how many miles were sailed each day? About _____

(unit)

8. Tutunendo, Colombia has the greatest annual rainfall in the world—about 464 inches per year. On average, about how many inches is that per month (to the nearest whole number)? About _____

(unit)

9. Tour buses at the zoo leave when every seat is occupied. Each bus holds 29 people. On Saturday, 1,827 people took tour buses. How many tour buses were filled? _____

(unit)

Try This

10. The diameter of the planet Neptune is about 30,600 miles. Pluto's diameter is about $\frac{1}{21}$ that of Neptune. About how many miles is the diameter of Pluto? About _____

(unit)

LESSON 2·8 Estimating and Calculating Quotients

For each problem, follow the steps below. Show your work on a separate sheet of paper or a computation grid.

◆ Estimate the quotient by using numbers that are "close" to the numbers given and that are easy to divide. Write your estimate. Then write a number sentence showing how you estimated.

◆ Ignore any decimal points. Divide as if the numbers were whole numbers.

◆ Use your estimate to insert a decimal point in the final answer.

1. 16)54.4   Estimate _____

How I estimated

_____

Answer _____

2. 12)206.4   Estimate _____

How I estimated

_____

Answer _____

3. $37.10 ÷ 14   Estimate _____

How I estimated

_____

Answer _____

4. 39.27 ÷ 21   Estimate _____

How I estimated

_____

Answer _____

5. 32)120   Estimate _____

How I estimated

_____

Answer _____

6. 28)232.4   Estimate _____

How I estimated

_____

Answer _____

LESSON
2·8   # Whole-Number Division with Decimal Answers

---

1. Seven people had lunch. The bill was $29. Divide $29 into 7 equal shares. About how much is 1 share, in dollars and cents?

    Answer _____

2. Find 8 ÷ 3. Give the answer as a decimal with 1 digit after the decimal point.

    Answer _____

---

3. Find 114 / 40. Give the answer as a decimal with 2 digits after the decimal point.

    Answer _____

4. A board 125 centimeters long is cut into 12 pieces of equal length. What is the length of each piece, to the nearest tenth of a centimeter?

    Answer _____

---

5. Find 26 / 19. Give the answer as a decimal with 2 digits after the decimal point.

    Answer _____

6. A 2-meter ribbon is cut into 6 pieces of equal size. What is the length of each piece, to the nearest tenth of a centimeter? *Hint:* How many centimeters are in 1 meter?

    Answer _____

LESSON 2·8 Math Boxes

1. Estimate each sum or difference. Then solve.

   a. 5.6 − 2.983   Estimate _____

   5.6 − 2.983 = _____

   b. 5.83 + 6.025 + 12.7   Estimate _____

   5.83 + 6.025 + 12.7 = _____

2. 43.2 / 16   Estimate _____

Use any division algorithm to solve 43.2 / 16. Show your work.

43.2 / 16 = _____

3. Multiply mentally.

   a. $27.8 * \frac{1}{10}$ = _____

   b. 3.0078 * 100 = _____

   c. $45.6 * \frac{1}{100}$ = _____

   d. 0.001 * 10,000 = _____

4. The hourly wages of 5 employees are $9.25, $10.50, $11.60, $11.75, and $12.20.

Explain how the median and mean would change if $9.25 were removed from the set of wages.

_____

_____

_____

5. Write the number pair for each of the following points shown on the coordinate grid.

   A: (_____,_____)

   B: (_____,_____)

   C: (_____,_____)

   D: (_____,_____)

   E: (_____,_____)

   F: (_____,_____)

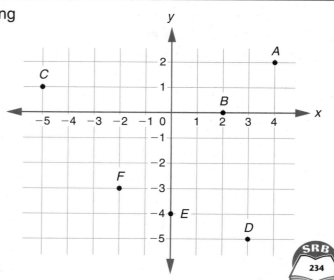

LESSON 2·8 Interpreting Remainders

Use any division algorithm to find each quotient. Then use the information in the problem to make sense of the remainder.

1. Mr. Jacobson has a can of Mancala beans. He asks Bill to divide the beans into sets of 48 beans each. There are 389 beans in the can. How many sets of 48 can Bill make?

   _____

2. Sharon is organizing her CD collection. She has 68 CDs in all. Each CD rack holds 25 CDs. How many racks does she need to hold her entire collection?

   _____

3. Jesse is making cookies for the class bake sale. He plans to charge 25 cents per cookie. He wants to make at least 342 cookies (one for each student in the school). Each batch makes about 25 cookies. How many batches should he make?

   _____

4. Mrs. Hanley's sixth-grade class is putting on a play for the rest of the school. They will set up rows of chairs in the gymnasium. If there are 21 chairs in each row, how many rows are needed for the 297 people who are expected to come?

   _____

5. Nestor's favorite sport is baseball. He went to 18 games last year. Nestor usually sits in the upper boxes, where tickets cost $15. This year, he's saved $238 for tickets. How many tickets will he be able to buy?

   _____

6. Regina, Ruth, Stanley, and Sam sold lemonade. They charged $0.20 per glass. They collected a total of $29.60. They paid $8.00 for supplies and evenly split the money that was left. How much did each of them receive?

   _____

LESSON 2·9 Positive and Negative Powers of 10

Some Powers of 10								
10^9	10^6	10^3	10^1	10^0	10^{-1}	10^{-3}	10^{-6}	10^{-9}
1,000,000,000	1,000,000	1,000	10	1	0.1	0.001	0.000001	0.000000001
1 billion	1 million	1 thousand	1 ten	one	1 tenth	1 thousandth	1 millionth	1 billionth

Use the table of powers and your knowledge of place value to write the following numbers in exponential notation.

1. 10,000 _____

2. 0.00001 _____

3. 10,000,000 _____

4. 0.01 _____

Write these powers of 10 in standard notation.

5. 10^{10} _____

6. 10^{-4} _____

7. 10^{-12} _____

8. 10^5 _____

In Lesson 2-4, you multiplied decimal numbers by powers of 10.

Use a power-of-10 strategy to find the following products. If you need to, look back at journal pages 56 and 57.

9. $9.65 * 10^3 =$ _____

10. $1.75 * 10^6 =$ _____

11. $4.2 * 10^{-2} =$ _____

12. $8.1 * 10^{-5} =$ _____

13. $5.748 * 10^0 =$ _____

14. $9.06 * 10^{-4} =$ _____

Write the missing power of 10.

15. $27,000 = 2.7 *$ _____

16. $538,000,000 = 5.38 *$ _____

17. $0.0003 = 3.0 *$ _____

18. $0.00000967 = 9.67 *$ _____

19. $845,603 = 8.45603 *$ _____

20. $0.00340051 = 3.40051 *$ _____

LESSON 2·9 Translating Standard & Scientific Notation

Scientific notation is used in the following facts. Rewrite each number in standard notation. Study the example.

Example: The number of taste buds in an
average human mouth is about $9 * 10^3$, or ____9,000____, taste buds.

1. The number of hairs on an average
   human body is about $5 * 10^6$, or _____, hairs.

2. The width of a hair is between
   $2 * 10^{-7}$ and $3 * 10^{-7}$, or _____ and _____, inches.

3. The length of a grain of salt is about $1 * 10^{-5}$, or _____, meter.

4. The time for a flea to take
   off on a vertical jump is $7.9 * 10^{-5}$, or _____, second.

Standard notation is used in the following facts. Rewrite each number in scientific notation. Study the example.

Example: The approximate cost of fuel for a
space shuttle mission is $2,000,000, or ____$2 * 10^6$____ dollars.

5. A Stradivarius violin in perfect
   condition sells for approximately $3,000,000, or _____ dollars.

6. The average growth rate of a child
   between birth and the age of 18 years is 0.007, or _____, inch per day.

7. The weight of a bee's brain
   is approximately 0.00004, or _____, ounce.

Try This

8. A CD (compact disc) holds about $6 * 10^9$ bits of information. This equals about $4 * 10^5$
   pages of text. A person's memory can hold about $1 * 10^{11}$ bits of information.

   a. About how many CDs' worth of
      information can one person's memory hold? _____
      (unit)

   b. About how many pages of text
      can one person's memory hold? _____
      (unit)

Sources: The Sizesaurus; The Compass in Your Nose and Other Astonishing Facts about Humans; and Everything Has Its Price

LESSON 2·9   **Math Boxes**

---

1. Use estimation to insert the decimal point in each product or quotient.

   a. 0.42 * 7 = 2 9 4

   b. 76.50 / 5 = 1 5 3 0

   c. 5.84 * 0.581 = 3 3 9 3 0 4

   d. 547.35 / 8.2 = 6 6 7 5

   SRB 37 42

---

2. Write each number in standard notation.

   a. 12.4 million _____

   b. 0.5 billion _____

   c. 5.3 trillion _____

   d. 0.75 million _____

---

3. Write each number in scientific notation.

   a. 7,000 _____

   b. 0.0008 _____

   c. 250,000,000 _____

   d. 0.0000395 _____

   SRB 7 8

---

4. The mayor of Calculus City claimed that the city's population doubled from 2000 to 2006.

   What would the population in 2006 have to be to make the mayor's claim true?

   _____

   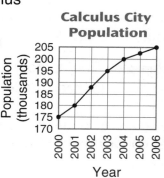
   Calculus City Population

   SRB 140

---

5. Which graph best represents the number of inches an average adult grows in a year? Circle the best answer.

   A.

   B.

   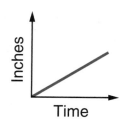

   C.

   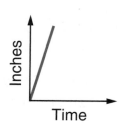

   SRB 140

74

LESSON 2·10 Exponential Notation and the Power Key

1. Complete the following table.

Words	Exponential Notation	Base	Exponent	Repeated Factors	Standard Notation
three to the fifth power	3^5	3	5	3 * 3 * 3 * 3 * 3	243
	7^4			7 * 7 * 7 * 7	
twenty to the third power					
				5 * 5 * 5 * 5 * 5 * 5	
ten to the negative fourth power	10^{-4}				

2. Complete the following table. Two possible key sequences are shown below.

Words	Exponential Notation	Base	Exponent	Calculator Key Sequence	Standard Notation
eight to the seventh power	8^7	8	7		2,097,152
ten to the sixth power					
	9^4				
two to the twenty-fifth power					
	10^{-3}			10 $\boxed{\wedge}$ $\boxed{(-)}$ 3 $\boxed{\text{Enter}}$ 10 $\boxed{x^y}$ 3 $\boxed{+/-}$ $\boxed{=}$	
five to the negative second power					

Math Boxes

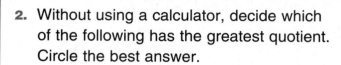

1. Using each of the digits 8, 4, 3, 2, 1, and 0 exactly once, make 2 decimal numbers whose difference is between 50 and 52.

_____ − _____ = _____

SRB
31 32

2. Without using a calculator, decide which of the following has the greatest quotient. Circle the best answer.

A. 23 / 0.04   **B.** 23 / 0.4

C. 23 / 4   **D.** 23 / 40

SRB
42

3. Use your calculator to help you convert the numbers written in exponential notation to standard notation.

a. $2^8 = $ _____

b. $15^5 = $ _____

c. $3^{-4} = $ _____

d. $2^{-3} = $ _____

SRB
285 286

4. In 1 week, the hours worked by the 6 cashiers at *Buy and Fly* were 20, 20, 29, 48, 52, and 56.

Would using the mean or the median number of hours encourage a person who wants to work at least 35 hours per week to apply for a job at *Buy and Fly?* Explain.

_____

_____

_____

_____

SRB
136 137

5. Write the number pair for each of the following points on the line graph.

P: (_____, _____)

Q: (_____, _____)

R: (_____, _____)

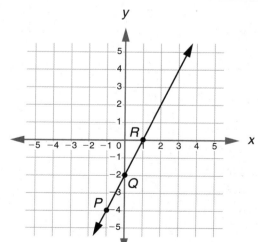

SRB
234

LESSON 2·10 — Multiplying and Dividing Decimals

1. A ticket for the school play costs $7.50. If 846 tickets are sold, what is the total for ticket sales?

   Estimate _____

   Answer _____

2. Brooke earns $450 per week before taxes. If she works 40 hours each week, what is Brooke's hourly rate of pay?

   Estimate _____

   Answer _____

3. You and 11 team members go out for lunch. The total, including tip, comes to $42.75. If you split the bill evenly, about how much should each person pay?

   Estimate _____

   Answer _____

4. In 1967, a 30-second commercial during the Super Bowl cost $42,000. In 2004, the cost was $2.5 million. About how many times as great was the cost in 2004 as in 1967?

   About _____

The table below shows customary and metric unit equivalents. Only the equivalent for inches and centimeters is exact. All the other equivalents are approximate.

Metric and Customary Unit Equivalents		
Length	**Mass/Weight**	**Capacity**
1 in. = 2.54 cm	1 oz is about 28.35 g	1 L is about 1.06 qt
1 m is about 3.28 ft	1 kg is about 2.2 lb	1 gal is about 3.79 L

Complete.

5. 76.2 cm = _____ in.

6. 500 m is about _____ ft.

7. 55 lb is about _____ kg.

8. 5 gal is about _____ L.

Scientific Notation on a Calculator

Use your calculator to help you fill in Column 2 first. Then fill in Columns 3 and 4 without using your calculator.

Exponential Notation	Calculator Display	Scientific Notation	Number-and-Word Notation
1. $1,000,000^2$		$1 * 10^{12}$	1 trillion
2. $10,000^3$			
3. $2,000,000^2$		$4 * 10^{12}$	
4. $20,000^3$			
5. $3,000,000^2$			
6. $30,000^3$			

Rewrite each of the following numbers in standard notation. Enter each number into your calculator and press the (Enter) key. Then write each number in scientific notation.

	Standard Notation	Scientific Notation
7. 456 million	_____	_____
8. 3.2 trillion	_____	_____
9. 23.4 billion	_____	_____
10. 78 trillionths	_____	_____

Solve each problem. Write each answer in standard and scientific notations.

11. $3,200,000 * 145,000$ _____ _____

12. 10 billion / 2.5 million _____ _____

13. $(5 * 10^4) - 10^2$ _____ _____

14. $10^2 + (3 * 10^3)$ _____ _____

15. $(8 * 10^{-1}) - 0.3$ _____ _____

16. $(4.1 * 10^8) - (3.6 * 10^6)$ _____ _____

17. $(17 * 10^{12}) / (4.25 * 10^6)$ _____ _____

18. $(6 * 10^{11}) + (7 * 10^9)$ _____ _____

Scientific Notation on a Calculator *continued*

Solve the number stories. Use a calculator and scientific notation to help you.

19. There are approximately 40 million students in the United States in Kindergarten through eighth grade. If, on average, each student goes to school for about 1,000 hours per year, about how many total hours do all of the students in Grades K–8 spend in school each year?

Answer _____

Number Model _____

20. The distance from Earth to the Sun is about 93 million miles. The distance from Jupiter to the Sun is about 483.6 million miles. About how many times as great is the distance from the Sun to Jupiter as the Sun to Earth?

Answer _____

Number Model _____

21. Light travels at a speed of about 186,300 miles per second.

   a. About how many miles does light travel in 1 minute?

      Answer _____

      Number Model _____

   b. About how many miles does light travel in 1 hour?

      Answer _____

      Number Model _____

   c. About how many miles does light travel in 1 day?

      Answer _____

      Number Model _____

22. The average distance from the Sun to Earth is about $9.3 * 10^7$ miles and from the Sun to Mars is about $1.4 * 10^8$ miles. What is the average distance from Earth to Mars?

Answer _____

Number Model _____

Math Boxes

1. Use estimation to insert the decimal point in each product or quotient.

   a. $0.62 * 9 = 5\ 5\ 8$

   b. $701.2 / 8 = 8\ 7\ 6\ 5$

   c. $11.47 * 0.843 = 9\ 6\ 6\ 9\ 2\ 1$

   d. $763.5675 / 9.15 = 8\ 3\ 4\ 5$

37 42

2. Convert the following numbers in standard notation to number-and-word notation.

   a. 4,600,000 _____

   b. 3,700,000,000 _____

   c. 8,340,000 _____

   d. 955,000 _____

3. Write each number in standard notation.

   a. $4.006 * 10^{-3}$ _____

   b. $8.0713 * 10^{5}$ _____

   c. $5 * 10^{-4}$ _____

   d. $9.22 * 10^{0}$ _____

7 8

4. A marketing company used the graph at the right to support their claim that candy sales have doubled.

No-Carb Candy Sales

Explain why the graph is misleading.

_____

_____

_____

_____

_____

138

5. Which graph most likely represents the number of cars driven through the car wash? Circle the best answer.

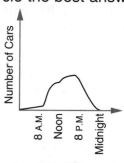

A.

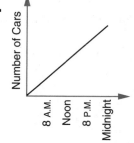

B.

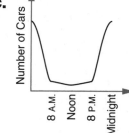

C.

140

Math Boxes

1. a. Plot the following points on the grid.

    A: (3,4)   *B:* (1,0)   *C:* (4,−3)

    D: (−3,−2)   *E:* (−1,3)

b. Draw line segments connecting points
A to *B, B* to *C, C* to *D, D* to *E,* and *E* to *A.*

c. What kind of polygon is this?

_____

d. Is it a convex or concave polygon?

_____

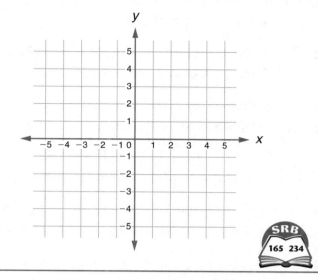

SRB
165 234

---

2. The area of a triangle can be found by
using the formula $A = \frac{1}{2} * (b * h)$, where
A is the area, *b* is the length of the base,
and *h* is the height. Find the area of the
triangle shown.

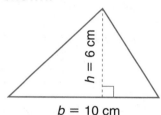

h = 6 cm

b = 10 cm

$A =$ _____ cm^2

SRB
217

---

3. Complete the "What's My Rule?" table.
Rule: Multiply by 1,000.

in	out
	500.00
	160
0.08	
0.002	
0.0007	

SRB
35 253

---

4. Reshi's brother was 8 years old when
Reshi was born.

a. If Reshi is now 5, how old is his brother?

_____

b. Suppose Reshi's brother is now twice
as old as Reshi. How old is Reshi?

_____

SRB
240

---

5. Plot each number on the number line and
write the letter label for the point.

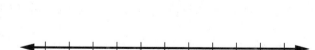

−5   0   5

  A: 1   *B:* −1

  C: 5   *D:* −5

SRB
99

LESSON 3·1 Patterns and Variables

Study the number sentences at the right. All three sentences show the same **general pattern.**

> $10\% \text{ of } 50 = \frac{10}{100} * 50$
>
> $10\% \text{ of } 200 = \frac{10}{100} * 200$
>
> $10\% \text{ of } 8 = \frac{10}{100} * 8$

◆ This general pattern may be described in words: To find 10% of a number, multiply the number by $\frac{10}{100}$ (or 0.10, or $\frac{1}{10}$).

◆ The pattern may also be described by a number sentence that contains a variable: $10\% \text{ of } n = \frac{10}{100} * n.$

A **variable** is a symbol, such as *n, x, A,* or ☐. A variable can stand for any one of many possible numeric values in a number sentence.

◆ Number sentences like $10\% \text{ of } 50 = \frac{10}{100} * 50$ and $10\% \text{ of } 200 = \frac{10}{100} * 200$ are examples, or **special cases,** for the general pattern described by $10\% \text{ of } n = \frac{10}{100} * n.$

To write a special case for a general pattern, replace the variable with a number.

Example:

General pattern $10\% \text{ of } \boldsymbol{n} = \frac{10}{100} * \boldsymbol{n}$

Special case $10\% \text{ of } \boldsymbol{35} = \frac{10}{100} * \boldsymbol{35}$

---

1. Here are 3 special cases for a general pattern.

   $\frac{10}{10} = 1$ $\frac{725}{725} = 1$ $\frac{\frac{1}{2}}{\frac{1}{2}} = 1$

   a. Describe the pattern in words.

   _____

   b. Give 2 other special cases for the pattern.

   _____ _____

2. Here are 3 special cases for another general pattern. $15 + (-15) = 0$ $3 + (-3) = 0$ $\frac{1}{4} + (-\frac{1}{4}) = 0$

   a. Describe the pattern in words.

   _____

   b. Give 2 other special cases for the pattern.

   _____ _____

LESSON 3·1 **Patterns and Variables** *continued*

SRB 103

3. A spider has 8 legs. The general pattern is: *s* spiders have *s* * 8 legs.
Write 2 special cases for the general pattern.

a. _____ b. _____

4. Study the following special cases for a general pattern.

The value of 6 quarters is $\frac{6}{4}$ dollars.

The value of 10 quarters is $\frac{10}{4}$ dollars.

The value of 33 quarters is $\frac{33}{4}$ dollars.

a. Describe the general pattern in words.

_____

b. Give 2 other special cases for the pattern.

_____

_____

Write 3 special cases for each general pattern.

5. $p + p = 2 * p$

_____

_____

_____

6. $c * \frac{1}{c} = 1$

_____

_____

_____

7. $p + p + (3 * p) = 5 * p$

_____

_____

_____

8. $s^2 + s = (s + 1) * s$

_____

_____

_____

LESSON 3·1 Writing General Patterns

Following is a method for finding the general pattern for a group
of special cases.

Example: Write the general pattern for the special cases at the right.

$$8 / 1 = 8$$

$$12.5 / 1 = 12.5$$

$$0.3 / 1 = 0.3$$

Solution Strategy

Step 1 Write everything that is the same for all of the special cases.

Use blanks for the parts that change.

_____ / 1 = _____

Each special case has division by 1 and an equal sign.

Step 2 Fill in the blanks. Each special case has a different number, but the number is
the same for both blanks, so use the same variable in both blanks.

Possible solutions: N / 1 = N , or x / 1 = x , or $\square$ / 1 = $\square$

Write a general pattern for each group of 3 special cases.

1. $18 * 1 = 18$

$2.75 * 1 = 2.75$

$\frac{6}{10} * 1 = \frac{6}{10}$ General pattern _____

2. $6 * 0 = 0$

$\frac{1}{2} * 0 = 0$

$78.7 * 0 = 0$ General pattern _____

3. 1 cat has $1 * 4$ legs.

2 cats have $2 * 4$ legs.

5 cats have $5 * 4$ legs.   General pattern _____

4. $6 * 6 = 6^2$

$\frac{1}{2} * \frac{1}{2} = \left(\frac{1}{2}\right)^2$

$0.7 * 0.7 = (0.7)^2$ General pattern _____

LESSON 3·1 Math Boxes

1. Complete the "What's My Rule?" table.

Rule: Subtract 1.32

in	out
8	
2.15	
1.8	
	3.57
	6.01

32 33
253

2. Write 3 special cases for the general pattern.

$a * (b + c) = (a * b) + (a * c)$

_____

_____

_____

103

3. Write each number in number-and-word notation.

a. 200,000

_____

b. 16,900,000,000

_____

c. 58,400,000,000,000

_____

4. Divide.

$6\overline{)93.6}$

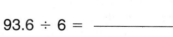

$93.6 \div 6 =$ _____

42–45

5. Add or subtract.

a. $\frac{1}{7} + \frac{4}{7} =$ _____

b. $\frac{7}{9} - \frac{2}{9} =$ _____

c. $\frac{3}{11} + \frac{7}{11} =$ _____

d. $1 - \frac{5}{6} =$ _____

83

6. Compare each pair of fractions. Write < or >.

a. $\frac{3}{10}$ _____ $\frac{5}{10}$

b. $\frac{83}{100}$ _____ $\frac{81}{100}$

c. $\frac{4}{15}$ _____ $\frac{5}{15}$

d. $\frac{19}{20}$ _____ $\frac{20}{20}$

75

LESSON
3·2
General Patterns with Two Variables

Math Message

The general pattern shown uses 2 variables.
They are a and b.

To write a special case for the general pattern:

◆ Replace the variable a with any number.

◆ Replace the variable b with any number.

Notice that in the third special case, the variables
a and b have been replaced by the same number.

General Pattern

$$a * (b - 1) = a * b - a$$

Special Cases

$$5 * (4 - 1) = 5 * 4 - 5$$

$$72 * (13 - 1) = 72 * 13 - 72$$

$$6 * (6 - 1) = 6 * 6 - 6$$

Write three special cases for each general pattern.

1. $x * y = y * x$

_____

_____

_____

2. $p * \frac{n}{n} = p$

_____

_____

_____

3. $x * 0 = y * 0$

_____

_____

_____

4. $\frac{a}{b} * \frac{b}{a} = 1$ (a and b are not 0.)

_____

_____

_____

5. $a + a + b = 2 * (a + b) - b$

_____

_____

_____

6. $(r + s) + (5 - s) = r + 5$

_____

_____

_____

LESSON 3·2 General Patterns with Two Variables *continued*

Example: Write a general pattern with 2 variables for the special cases at the right.

$$7 + 3 = 3 + 7$$

$$\frac{1}{2} + \frac{3}{2} = \frac{3}{2} + \frac{1}{2}$$

$$4 + (-2) = -2 + 4$$

Solution Strategy

Step 1 Write everything that is the same for all of the special cases. Use blanks for the parts that change.

_____ + _____ = _____ + _____

Each special case has two additions and an equal sign.

Step 2 Each special case has 2 different numbers. Use different variables (letters or other symbols) for the numbers that vary. Write them on the blanks.

__a__ + __b__ = __b__ + __a__

Step 3 Check that the special cases given fit the general pattern.

Write a number sentence with 2 variables for each general pattern.

1. $4 * \frac{2}{7} = 2 * \frac{4}{7}$

$10 * \frac{2}{3} = 2 * \frac{10}{3}$

$29 * \frac{2}{8} = 2 * \frac{29}{8}$

General pattern _____

2. $(5 * 2) + (5 * 6) = 5 * (2 + 6)$

$(5 * 4) + (5 * 1) = 5 * (4 + 1)$

$(5 * 2) + (5 * 100) = 5 * (2 + 100)$

General pattern _____

3. Write a general pattern using variables.
Let d = number of dogs.
Let b = number of birds.

3 dogs and 5 birds have
$(3 * 4) + (5 * 2)$ legs.

5 dogs and 9 birds have
$(5 * 4) + (9 * 2)$ legs.

17 dogs and 6 birds have
$(17 * 4) + (6 * 2)$ legs.

_____

4. For the general pattern $x^2 * y^2 = (x * y)^2$, write the special case.

a. $x = 4$ and $y = 5$

_____

b. $x = 10$ and $y = 10$

_____

c. Is the general pattern true, no matter which numbers you use? _____

LESSON 3·2 **Math Boxes**

1. Three special cases of a pattern are given below. Using 2 variables, write a number sentence to describe the general pattern.

   $4 * (2 + 3) = (4 * 2) + (4 * 3)$

   $4 * (1 + 9) = (4 * 1) + (4 * 9)$

   $4 * (5 + 7) = (4 * 5) + (4 * 7)$

   General pattern:

   _____

   SRB
   103

2. Evaluate each expression by substituting 5 for t.

   a. $12 - t$ _____

   b. $2.5t$ _____

   c. $\dfrac{t}{5}$ _____

   d. $4t - 7$ _____

   SRB
   242 243

3. Write in standard notation.

   a. 5^{-4} _____

   b. 4^{-1} _____

   c. $5 * 10^{-3}$ _____

   d. $6 * 10^{-2}$ _____

   SRB
   8

4. The heights of 5 starting players on a basketball team, rounded to the nearest centimeter, are:
   173 cm, 191 cm, 178 cm, 185 cm, 188 cm.

   Find the median height.

   median _____

   SRB
   136

5. Mark and label each fraction on the number line.

   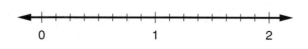

   a. Point $A = \dfrac{1}{4}$ b. Point $B = 1\dfrac{1}{8}$

   c. Point $C = 1\dfrac{3}{8}$ d. Point $D = \dfrac{3}{2}$

   SRB
   68

6. List all the factors of each number.

   30 _____

   24 _____

   Name the greatest common factor (GCF) of 30 and 24.

   _____

   SRB
   80

Date _____   Time _____

Algebraic Expressions

Example: The hardcover edition of a book costs $20 more than the paperback edition.

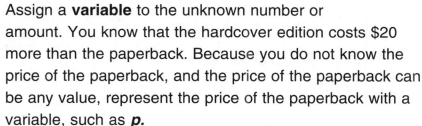

Step 1 Assign a **variable** to the unknown number or amount. You know that the hardcover edition costs $20 more than the paperback. Because you do not know the price of the paperback, and the price of the paperback can be any value, represent the price of the paperback with a variable, such as *p.*

Let p = the price of the paperback edition.

Step 2 Write an **algebraic expression.**
Since p represents the price of the paperback edition, $p + \$20$ is the algebraic expression that represents the price of the hardcover edition.

Step 3 **Evaluate** the algebraic expression.
You can evaluate the algebraic expression $p + \$20$ by assigning a value to the variable p. For example, if $p = \$10.95,$ then you would evaluate $p + \$20$ as $\$10.95 + \$20,$ or $30.95. So, if the price of the paperback edition is $10.95, then the price of the hardcover edition is $30.95.

Complete each of the following statements with an algebraic expression, using the suggested variable.

1. If a large pizza costs $4.25 more than a small pizza,

   then a large pizza costs _____ dollars.

A small pizza costs d dollars.

large pizza

2. **a.** If Boris's hamster is 12 months older than his goldfish,

   then his goldfish is _____ months old.

   b. Evaluate your expression. If Boris's hamster is 21 months old, how old is his goldfish?

   _____ months old

Boris's hamster is h months old.

Boris's goldfish

LESSON 3·3   Algebraic Expressions *continued*

Complete each of the following statements with an algebraic expression,
using the suggested variable.

3. a. The weight of 5 bags of candy is _____ pounds.

A bag of candy weighs
p pounds.

b. If every member of your class had a bag of candy, how
many pounds of candy would there be?

_____

4. If the whale dives 85 feet, it will be at a depth of

_____ feet.

The whale is at a depth
of m feet.

5. a. During lunch, the cafeteria is divided into 3 parts,
with each part having an equal floor area. Each part

has an area of _____ ft^2.

b. If the area of the cafeteria floor is 2,400 square feet,

what is the area of each of the 3 parts? _____
(unit)

The cafeteria floor has an
area of A ft^2.

Try This

6. The charge for a book that
is d days overdue is

_____ cents.

A library charges 5 cents for
each day a book is overdue,
plus an additional 10-cent
service charge.

The Meaning of Cave Art
by
Nancy Neanderthal

7. a. If James spends $\frac{5}{6}$ of his weekly
allowance seeing a movie, he has

_____ dollars of his allowance left.

b. If a movie ticket costs $7.50,
what is James's weekly allowance? _____

James's weekly
allowance is x dollars.

LESSON 3·3 Math Boxes

---

1. Complete the "What's My Rule?" table.

Rule: Subtract 0.58

in	out
7	
1.09	
0.9	
	4.62
	3.15

SRB 32 33 253

---

2. Write 3 special cases for the general pattern.

$x * (y - z) = (x * y) - (x * z)$

_____

_____

_____

SRB 103

---

3. Which is the number-and-word notation for 3,856,000,000?

Choose the best answer.

- ◯ 38.56 million
- ◯ 3.856 million
- ◯ 3.856 billion
- ◯ 3.856 trillion

---

4. Divide $58 into 14 equal shares.

Estimate _____

About how much is one share, in dollars and cents?

_____

SRB 42–45

---

5. Add or subtract.

a. $\frac{8}{9} - \frac{4}{9} =$ _____

b. $\frac{2}{15} + \frac{4}{15} =$ _____

c. $1 - \frac{5}{8} =$ _____

d. $\frac{9}{10} + \frac{3}{10} + \frac{1}{10} =$ _____

SRB 83

---

6. Compare each pair of fractions. Write < or >.

a. $\frac{3}{4}$ _____ $\frac{3}{7}$

b. $\frac{4}{9}$ _____ $\frac{4}{5}$

c. $\frac{10}{3}$ _____ $\frac{10}{2}$

d. $\frac{3}{5}$ _____ $\frac{3}{3}$

SRB 75

LESSON 3·3 Division Practice

For Problems 1–4:

◆ Estimate the quotient. Write a number sentence to show how you estimated.

◆ Divide. Give the answer to two decimal places. If there are not enough decimal places in the dividend (the number being divided), add on as many zeros as necessary. Use your estimate to place the decimal point in your answer.

1. $8\overline{)983}$ Estimate _____

How I estimated _____

Answer _____

2. $12\overline{)437}$ Estimate _____

How I estimated _____

Answer _____

3. $46\overline{)728}$ Estimate _____

How I estimated _____

Answer _____

4. $11\overline{)652}$ Estimate _____

How I estimated _____

Answer _____

For Problems 5 and 6:

◆ Estimate the quotient. Write a number sentence to show how you estimated.

◆ Ignore the decimal point and divide. *Disregard any remainder.*

◆ Use your estimate to place the decimal point in your answer.

5. $5\overline{)315.8}$ Estimate _____

How I estimated _____

Answer _____

6. $8\overline{)204.6}$ Estimate _____

How I estimated _____

Answer _____

LESSON 3·4   # Using Formulas

Example: Find the perimeter (*P*) of the rectangle shown below.

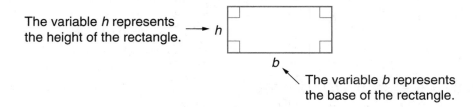

The variable *h* represents the height of the rectangle.

The variable *b* represents the base of the rectangle.

◆ One way to find the perimeter of the rectangle is to find the sum of the lengths of the sides.

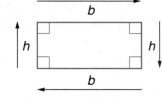

The variable *P* represents the perimeter of the rectangle.   → $P = b + h + b + h$

◆ Another way is to find the distance halfway around the rectangle and multiply that distance by 2.

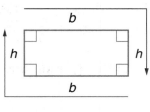

$$P = 2 * (b + h)$$

Use the formula $P = 2 * (b + h)$ to find the perimeter of a rectangle if $b = 8.5$ inches and $h = 4.5$ inches.

Solution Strategy

Estimate: $2 * (9 \text{ in.} + 5 \text{ in.}) = 2 * 14 \text{ in.} = 28 \text{ in.}$

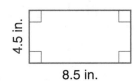

$P = 2 * (b + h)$

$\quad = 2 * (8.5 \text{ in.} + 4.5 \text{ in.})$

$\quad = 2 * (13 \text{ in.})$

$\quad = 26 \text{ in.}$

$P = 26 \text{ in.}$

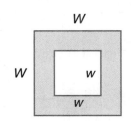

1. The area of the shaded region between the two squares in the diagram can be found by using the formula $A = W^2 - w^2$. Find the area of the shaded region if $W = 20$ inches and $w = 9$ inches.

Be careful! The capital *W* and the lowercase *w* stand for different lengths.

_____ in.^{2}

93

LESSON 3·4   **Evaluating Formulas**

1. The formula for the area of a parallelogram is $A = b * h$. You can use the formula for the area of a parallelogram to construct the formula for the area of a triangle.

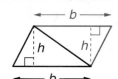

   Use the formula $A = \frac{1}{2} * b * h$ to find the area of a triangle with $b = 4$ cm and $h = 2.5$ cm. $A =$ _____ cm^2

2. You can use the formula $C = 2 * \pi * r$ to find the circumference of a circle, where C is the circumference and r is the radius.

   a. What is the circumference of a circle with a radius of 1 ft?

   b. What is the circumference of a circle with a radius of 12 in.?

   _____ ft

   _____ in.

3. You can use the formula $c = 1.1 * t$ to find the average number of calories a typical adult uses while lying in bed resting. The variable t represents the number of minutes a person is resting, and c is the number of calories used while resting.

   a. How many calories does a resting adult use in 12 minutes? _____

   (unit)

   b. How many calories does that adult use in 120 seconds? _____

   (unit)

4. The size of the screen on a television set or computer monitor is reported as its diagonal length. For example, the screen on a 17-inch monitor has a diagonal length of 17 inches.

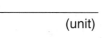

   The two formulas at the right show how the diagonal length d, the base length b, and the height h are related.

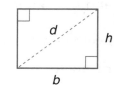

   $b = 0.8 * d$
   $h = 0.6 * d$

   Find the base length and the height of a 21-inch television screen.

   a. base _____

   (unit)

   b. height _____

   (unit)

LESSON 3·4 "What's My Rule?"

Complete each table according to the rule.

1. Rule: $y = x - 6.3$

x	y
37	
−15	
	16.5
0	−6.3
3	

2. Rule: $b = 4 / a$

a	b
2	
$\frac{1}{2}$	
5	
	1
	$\frac{1}{2}$

3. Rule: $n = 0.5 * m$

m	n
2	
10	
0.8	
	0.5
	3

4. The formula $F = (1.8 * C) + 32$ can be used to convert degrees Celsius to degrees Fahrenheit.

Rule: $F = (1.8 * C) + 32$

C	F
5°	
10°	
30°	
	32°
	33.8°

5. a. State in words the rule for the "What's My Rule?" table at the right.

_____

_____

b. Circle the formula that describes the rule.

$t - s = 0$ $t = s \div 5$ $s = t + 3t$

s	t
5	1
25	5
100	20
0	0
1	$\frac{1}{5}$

LESSON 3·4

Math Boxes

1. Three special cases of a pattern are given below. Using one variable, write a number sentence to describe the general pattern.

   $4 + 4 - 9 = (2 * 4) - 3^2$

   $8 + 8 - 9 = (2 * 8) - 3^2$

   $3.5 + 3.5 - 9 = (2 * 3.5) - 3^2$

   General pattern:

   _____

   SRB 103

2. Evaluate each expression by substituting 0.25 for x.

   a. $9 - x$ _____

   b. $100x$ _____

   c. $x / 10$ _____

   d. $4x - 12$ _____

   SRB 242 243

3. Write in standard notation.

   a. 2^{-3} _____

   b. 8^{-2} _____

   c. $2 * 10^{-3}$ _____

   d. $8 * 10^{-1}$ _____

   SRB 8

4. A company president claims that half of the company's employees earn more than $60,000 per year and half earn less.

   Which landmark best represents the situation described above?

   Choose the best answer.

   ⬭ range   ⬭ mode

   ⬭ mean   ⬭ median

   SRB 136

5. Mark and label each fraction on the number line.

   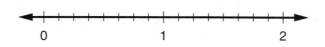

   0   1   2

   a. Point $A = \frac{1}{2}$ b. Point $B = \frac{9}{8}$

   c. Point $C = \frac{7}{8}$ d. Point $D = 1\frac{3}{4}$

   SRB 68

6. List all the factors of each number.

   45 _____

   15 _____

   Name the greatest common factor (GCF) of 45 and 15.

   _____

   SRB 80

LESSON 3·5

Math Boxes

---

1. Give 3 special cases for the general pattern $\frac{0}{k} = k - k$.

_____

_____

_____

SRB 103

---

2. Complete each of the following statements with an algebraic expression using the suggested variable.

   a. Trisha is t years old. If Kyle is 3 years younger than Trisha, then Kyle is

   _____ years old.

   b. Damien has d DVDs. If Nadia has half as many DVDs as Damien, then Nadia

   has _____ DVDs.

SRB 240

---

3. The time of day varies from time zone to time zone. The time difference between Newark, New Jersey, and Seattle, Washington, is given by the formula $n - s = 3$, where n stands for the time in Newark and s for the time in Seattle.

   a. If $s = 8$ P.M., $n =$ _____.

   b. If $n = 8$ P.M., $s =$ _____.

   c. If $s = 11$ P.M., $n =$ _____.

SRB 245 246

---

4. Write in standard notation.

   a. $2^4 =$ _____

   b. $3^4 =$ _____

   c. $5.2 * 10^3 =$ _____

   d. $6.4 * 10^{-5} =$ _____

SRB 8

---

5. Rename each fraction as a decimal.

   a. $\frac{7}{10} =$ _____

   b. $\frac{13}{50} =$ _____

   Rename each decimal as a fraction.

   c. $0.03 =$ _____

   d. $0.75 =$ _____

SRB 59 60

---

6. Rename each fraction as an equivalent fraction.

   a. $\frac{1}{4} =$ _____   **b.** $\frac{1}{2} =$ _____

   c. $\frac{1}{10} =$ _____   **d.** $\frac{6}{8} =$ _____

   e. $\frac{4}{5} =$ _____

SRB 73

LESSON 3·5

Representing Speed

SRB
253 254

Math Message

Ever since the early years of the automobile, police have been trying to catch speeding motorists. At first, officers were equipped with bicycles and stopwatches. They hid behind trees or rocks and came out to pursue speeders. An Englishman, Lord Montague of Beaulieu, received a ticket for driving 12 miles per hour.

Source: Beyond Belief!

"Eagle Eye" Gus Schalkman holds the record for ticket writing. He got his 135th car on his post near the Queensboro bridge on Aug. 14, 1929.

UPI - Bettman

1. At a speed of 12 miles per hour, about how many miles could Lord Montague travel in 1 minute? _____ mi

2. Tell how you solved this problem. _____

_____

3. Complete the table. Use the formula $d = 0.2 * t$.

   A car traveling at a speed of 12 miles per hour is traveling 0.2 mile per minute. You can use the following rule to calculate the distance traveled for any number of minutes.

   Distance traveled (*d*) = 0.2 mile per minute * number of minutes (*t*)

   You can also use the formula $d = 0.2 * t$ where *d* stands for the distance traveled in miles and *t* for the time traveled in minutes.

   For example, in 1 minute, the car will travel 0.2 mile (0.2 * 1). In 2 minutes, it will travel 0.4 mile (0.2 * 2).

Time (min) *t*	Distance (mi) 0.2 * *t*
1	0.2
2	0.4
3	
4	
5	
6	
7	
8	
9	
10	

Date _____   Time _____

LESSON 3·5 **Representing Speed** *continued*

4. Complete the graph using the table on page 98. Connect the points.

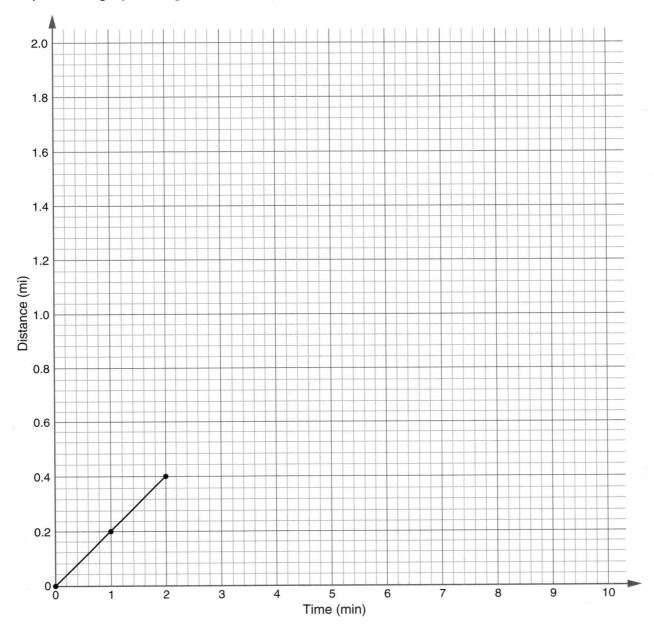

Use your graph to answer the following questions.

5. a. About how far would the car travel in $1\frac{1}{2}$ minutes?   About _____ (unit)

b. About how many miles would the car
travel in 5 minutes 24 seconds (about $5\frac{1}{2}$ minutes)?   About _____ (unit)

6. About how long would it take the car to travel 5 miles?   About _____ (unit)

99

LESSON 3·5 Representing Rates

SRB 253 254

Complete each table. Then graph the data. Connect the points.

1. Emma earns $6 per hour.

   Rule:
   Earnings = $6/hr
   ∗ number of hr worked

   Formula:
   $E = 6 * h$

Time (hr) h	Earnings ($) 6 * h
1	
2	
3	
	27
7	

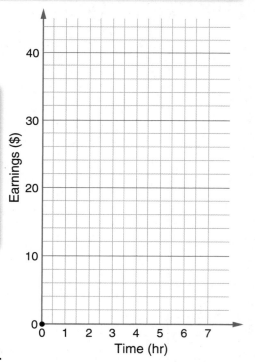

   a. Plot a point to show Emma's earnings for $5\frac{1}{2}$ hours.

   b. How much would she earn? _____

2. A young blue whale can gain as much as 300 pounds per day.

   Rule:
   Weight gained =
   300 lb/day
   ∗ number of days

   Formula:
   $W = 300 * t$

Time (days) t	Weight gained (lb) 300 * t
1	
2	
3	
	1,425
6	

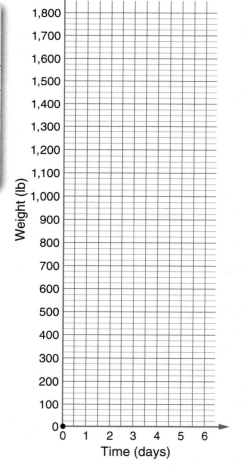

   a. Plot a point to show the number of pounds a young blue whale can gain in 36 hours.

   b. How many pounds is that? _____

   Source: Beyond Belief!

LESSON 3·5   **Representing Rates** *continued*

SRB
253 254

3. Chewy candies cost $1.50 a pound.

Rule:
Cost = $1.50/lb
* number of lb

Formula:
c = 1.50 * w

Weight (lb) w	Cost ($) 1.50 * w
1	
2	
3	
	15.00
12	

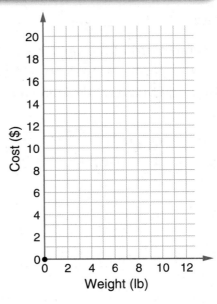

a. Plot a point to show the cost of 8 pounds.

b. How much would 8 pounds cost? _____

4. An average 11-year-old reads about 11 pages of text per day.

Rule:
Pages read =
11 pages/day
* number of days

Formula:
p = 11 * t

Time (days) t	Pages 11 * t
1	
2	
3	
	44
$5\frac{1}{2}$	

a. Plot a point to show how many pages of text an average 11-year-old reads in a week.

b. How many pages is that? _____

Source: Astounding Averages

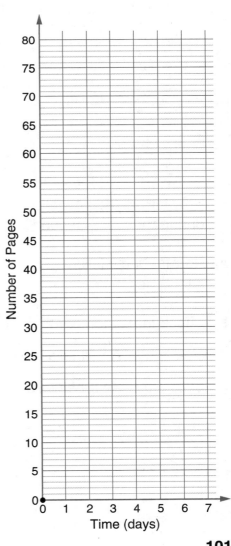

101

LESSON 3·6   Falling Objects

Math Message

The picture at the right was drawn from flash photographs of a falling golf ball. The time interval between flashes was $\frac{1}{20}$ second.

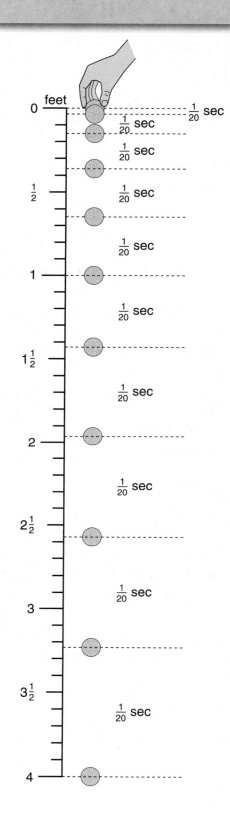

Elapsed Time (sec)	Total Distance Fallen (ft)
$\frac{1}{20}$	0.04
$\frac{2}{20}$	0.16
$\frac{3}{20}$	0.36
$\frac{4}{20}$	0.64
$\frac{5}{20}$	1.00
$\frac{6}{20}$	1.44
$\frac{7}{20}$	1.96
$\frac{8}{20}$	2.56
$\frac{9}{20}$	3.24
$\frac{10}{20}$	4.00

1. How far did the ball fall during the first $\frac{1}{4}$ second? _____

2. How far had it fallen after $\frac{1}{2}$ second? _____

3. Check the statement that you believe is true.

_____ A ball falls at a constant (even) speed.

_____ As a ball falls, it picks up speed.

LESSON 3·6 **Unit Prices and the Better Buy**

A rate is a comparison of two quantities with different units. If the comparison is to 1 unit, the rate is called a unit rate. You can use division to find a unit rate.

Example: Suppose you travel 215 miles in 4 hours.

$$\text{Rate} = \frac{215 \text{ mi}}{4 \text{ hr}} = 215 \text{ mi} \div 4 \text{ hr} \qquad \text{Unit rate} = \frac{53.75 \text{ mi}}{1 \text{ hr}}$$

Write each as a unit rate.

1. 810 kilometers on 45 liters of gasoline _____

2. $502 for 40 hours of work _____

3. 513 calories in 9 crackers _____

Many consumers decide which products to buy by comparing unit prices. A unit price is the cost per one unit of an item. Study the example below.

Example:

Shampoo X	Shampoo Y
$5.40 for an 8-oz bottle	**$8.34 for a 12-oz bottle**
Find the unit price.	Find the unit price.
$\text{Unit Price} = \frac{\text{Cost}}{\text{Unit}}$	$\text{Unit Price} = \frac{\text{Cost}}{\text{Unit}}$
$\text{Unit Price} = \frac{\$5.40}{8 \text{ oz}}$	$\text{Unit Price} = \frac{\$8.34}{12 \text{ oz}}$
$= 5.40 \div 8$	$= 8.34 \div 12$
$= 0.675$	$= 0.695$
$= \$0.68 \text{ per ounce}$	$= \$0.70 \text{ per ounce}$
Shampoo X costs $0.68/oz.	Shampoo Y costs $0.70/oz.

4. The item with the lower unit price is usually considered to be the better buy. Which shampoo is the better buy? Shampoo _____

5. Which is the better buy?

   a. $5.49 for a 12-oz box of cereal

      or

      $7.20 for a 16-oz box of cereal

   _____

   b. $3.99 for 6 cans of soup

      or

      $15.69 for 24 cans of soup

   _____

LESSON 3·6 · What Happens When an Object Is Dropped?

Galileo was an Italian physicist who lived from 1564 to 1642. His work led to the following rule for the distance traveled by a freely falling object.

$$\text{Distance traveled in feet} = 16 * \text{square of the elapsed time in seconds}$$

Written as a formula, the rule is $d = 16 * t * t$; or $d = 16 * t^2$, where d is the distance traveled by the object in feet and t is the time in seconds that has elapsed since the object started falling. For example, after 1 second, an object will have traveled 16 feet ($16 * 1 * 1$); after 2 seconds, it will have traveled 64 feet ($16 * 2 * 2$).

Galileo's formula really applies only in a vacuum, where there is no air resistance to slow an object's fall. However, it is a good approximation for the fall of a dense object, such as a bowling ball, over a fairly short distance.

1. The following table shows the approximate distance traveled by a freely falling object at 1-second intervals. Use Galileo's formula to complete the table.

Elapsed time (sec) t	0	1	2	3	4	5	6	7	8	9
Distance (ft) $16 * t^2$	0	16	64	144						

2. Graph the data from the table above onto the grid on page 105.

3. Use the graph to estimate the number of seconds it takes an object to fall 500 feet, ignoring air resistance.   About _____ seconds

Try This

4. a. About how many seconds would it take an object to fall 1 mile, ignoring air resistance? Use your calculator. (*Hint:* 1 mile = 5,280 feet)   About _____ seconds

   b. In the table at the right, evaluate Galileo's distance formula for at least two other values of t (time).

t		
$16 * t^2$		

Date _____   Time _____

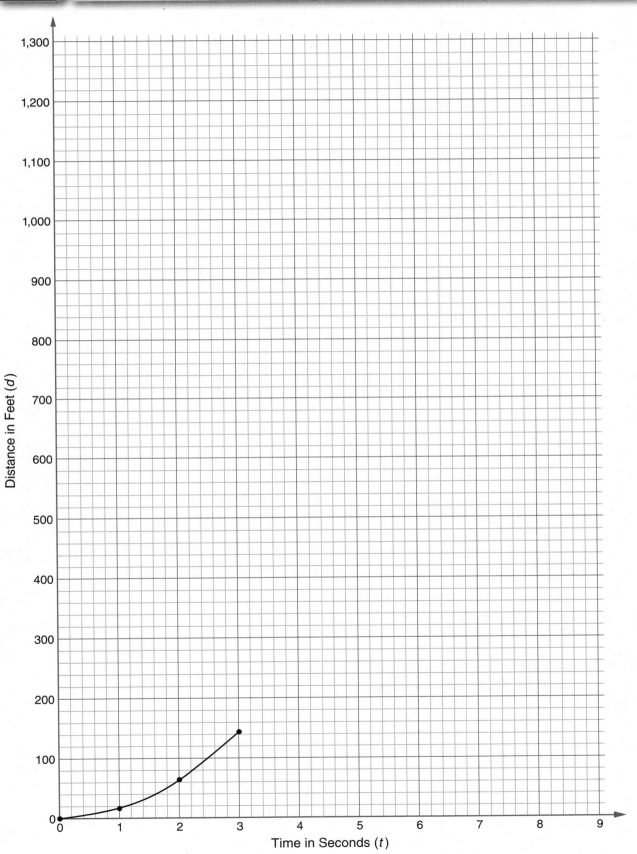

Distance in Feet (*d*)

Time in Seconds (*t*)

LESSON 3·6   When an Object Is Dropped *continued*

The speed of a freely falling object increases the longer it continues to fall. You can calculate the speed of a falling object at any given instant by using the rule

speed in feet per second = 32 * elapsed time in seconds, or by

using the formula $s = 32 * t$.

In the formula, s is the speed of the object in feet per second and t is the elapsed time in seconds since the object started falling.

For example, after 1 second, the object is traveling at a speed of about 32 feet per second (32 * 1); and after 2 seconds, at a speed of about 64 feet per second (32 * 2).

5. Use this formula to complete the table. Then graph the data in the table.

Elapsed time (sec) t	Speed (ft per sec) $32 * t$
1	32
2	64
3	
4	
5	
6	

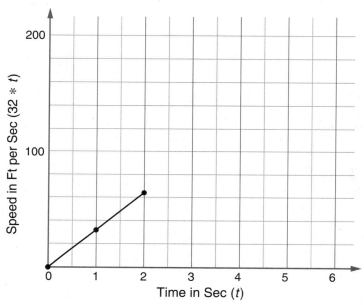

6. At about what speed would an object be traveling 18 seconds after it started falling?

About _____ ft per sec

7. If a freely falling object is traveling at a speed of 290 feet per second, about how long has it been falling?

About _____ sec

LESSON 3·6   **Math Boxes**

---

1. The area A of a circle can be found by using the formula $A = \pi * r * r$. Use $\pi = 3.14$.

   a. Find the area if $r = 3$ centimeters.

   _____ cm²

   b. Find the area if $d = 10$ inches.

   _____ in.²

   SRB 218

2. Give 3 special cases for the general pattern below.

   $(b * h) + 4 = 4 + (h * b)$

   _____

   _____

   _____

   SRB 103

---

3. Complete the table. Then graph the data and connect the plotted points.

   Samantha earns $9 for each yard she mows.

   Rule: Earnings = $9 * number of yards mowed

Number of yards (y)	Earnings ($) (9 * y)
0	0
1	
2	
3	
	45
6	

   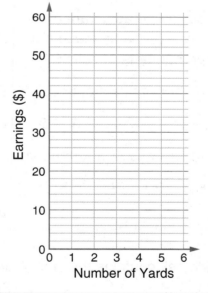

   SRB 254

---

4. Rename each decimal as a percent.

   a. 0.49 = _____

   b. 0.03 = _____

   c. 0.3 = _____

   d. 0.045 = _____

   SRB 59 60

5. List the first 6 multiples of each number.

   18 _____

   72 _____

   Name the least common multiple (LCM) of 18 and 72.

   _____

   SRB 78

107

LESSON 3·7 Variables and Formulas in Spreadsheets

A typical computer spreadsheet has **columns,** identified by letters, and **rows,** identified by numbers. Columns and rows intersect to form boxes called **cells.**

Each cell in a spreadsheet is named by the letter of the column and the number of the row it is in. For example, cell A1 is in column A, row 1. There is no space in the name between the letter and the number. Cells can contain text, numbers, or nothing at all.

Example: The spreadsheet at the right shows the number of hits made and runs scored by players on a softball team. The statistics are for the first 5 games played. As more games are played, the numbers will be updated.

Carl made 9 hits and scored 4 runs. His name is in cell A2, the number 9 is in cell B2, and the number 4 is in cell C2.

Think of a cell name as a variable. As the team plays more games, Carl will probably make more hits and score more runs, and the numbers in cells B2 and C2 will change.

	A	B	C
1	Player	Hits	Runs
2	Carl	9	4
3	Amala	5	2
4	Doug	1	0
5	Noreen	11	5
6	David	3	3
7	Annina	2	1
8	Ted	7	3
9	Raoul	12	7
10	Cheryl	3	0
11			
12	**Total**	**53**	**25**

The total number of hits, 53, in cell B12 is the following sum:

53 = 9 + 5 + 1 + 11 + 3 + 2 + 7 + 12 + 3

This is a special case of a formula that can be written using the cell names.

B12 = B2 + B3 + B4 + B5 + B6 + B7 + B8 + B9 + B10

1. **a.** What is in cell A5? _____   **b.** What is in cell B3? _____

2. **a.** Which cell contains the word Runs? _____

   b. Which cell contains the number 12? _____

   c. Which cell contains the total number of runs scored by all of the players? _____

3. Write a formula for calculating C12 that uses the cell names.

   _____

4. Suppose Raoul scored only 4 runs instead of 7. Which cell entries would change, and what would they change to?

   _____

LESSON 3·7

Spreadsheet Scramble Game Mats

	A	B	C	D	E	F
1						**Total**
2						
3						
4						
5	**Total**					

	A	B	C	D	E	F
1						**Total**
2						
3						
4						
5	**Total**					

	A	B	C	D	E	F
1						**Total**
2						
3						
4						
5	**Total**					

	A	B	C	D	E	F
1						**Total**
2						
3						
4						
5	**Total**					

	A	B	C	D	E	F
1						**Total**
2						
3						
4						
5	**Total**					

	A	B	C	D	E	F
1						**Total**
2						
3						
4						
5	**Total**					

	A	B	C	D	E	F
1						**Total**
2						
3						
4						
5	**Total**					

	A	B	C	D	E	F
1						**Total**
2						
3						
4						
5	**Total**					

LESSON 3·7   Preparing for Fraction Computation

Rename each mixed number as a fraction.

1. $1\frac{3}{4}$ _____

2. $3\frac{1}{3}$ _____

3. $10\frac{1}{10}$ _____

4. $12\frac{1}{8}$ _____

5. $2\frac{2}{5}$ _____

6. $4\frac{5}{6}$ _____

7. $9\frac{3}{7}$ _____

8. $1\frac{9}{16}$ _____

9. $6\frac{2}{3}$ _____

Rename each fraction as a mixed or whole number.

10. $\frac{8}{3}$ _____

11. $\frac{17}{4}$ _____

12. $\frac{22}{5}$ _____

13. $\frac{45}{9}$ _____

14. $\frac{9}{5}$ _____

15. $\frac{6}{3}$ _____

16. $\frac{37}{11}$ _____

17. $\frac{30}{9}$ _____

18. $\frac{66}{7}$ _____

List the factors of each number. Then find the largest factor that both numbers have in common.

19. 10 _____

15 _____

Greatest common factor _____

20. 8 _____

12 _____

Greatest common factor _____

List the first 6 multiples of each number. Then find the smallest multiple that both numbers have in common.

21. 8 _____

12 _____

Least common multiple _____

22. 3 _____

9 _____

Least common multiple _____

LESSON 3·7   **Math Boxes**

---

1. Give 3 special cases for the general pattern $\frac{a}{a} = 2 - \frac{a}{a}$.

_____

_____

_____

SRB
103

---

2. Which algebraic expression represents the following word phrase? Circle the best answer.

The quotient of y and 5

A. $\frac{y}{5}$

B. $y - 5$

C. $5y$

D. $5 - y$

SRB
240

---

3. The relationship between Blaire's and Katie's weekly allowance is expressed by the formula $b = 2 * k$, where b stands for Blaire's allowance and k for Katie's.

a. If $k = \$10.00$, $b =$ _____ .

b. If $b = \$30.00$, $k =$ _____ .

c. If $k = \$8.50$, $b =$ _____ .

SRB
245 246

---

4. Write in standard notation.

a. $5^3 =$ _____

b. $7^3 =$ _____

c. $0.348 * 10^7 =$ _____

d. $129 * 10^{-6} =$ _____

SRB
8

---

5. Rename each fraction as a decimal.

a. $\frac{1}{4} =$ _____

b. $\frac{7}{25} =$ _____

Rename each decimal as a fraction.

c. $0.80 =$ _____

d. $0.031 =$ _____

SRB
59 60

---

6. Rename each fraction as an equivalent fraction.

a. $\frac{1}{8} =$ _____   **b.** $\frac{3}{7} =$ _____

c. $\frac{8}{12} =$ _____   **d.** $\frac{4}{6} =$ _____

e. $\frac{6}{10} =$ _____

SRB
73

LESSON 3·8   Spreadsheet Practice

Math Message

Students are selling coupon books to raise money for the school band. A coupon book sells for $2.50. When the spreadsheet at the right is completed, it will show how many books these 4 students sold and how much money they collected.

	A	B	C
1	Student	Books	Money
2	Luigi	5	$12.50
3	Robin	16	
4	Akira	30	
5	Gloria	13	
6	**Total**		

1. Fill in cell B6 to show the total number of books sold.

2. Fill in cells C3, C4, and C5 to show how much money Robin, Akira, and Gloria collected.

3. Fill in cell C6 to show the total amount of money collected.

The spreadsheet below shows students' test scores for 2 different tests.

	A	B	C	D
1	Student	Test 1	Test 2	Average
2	Amy	85	90	87.5
3	David	70	86	
4	Amit	78	64	
5	Beth	65	81	

4. Calculate the remaining average test scores. Fill in the spreadsheet.

5. David's Test 1 score is in which cell? _____

6. The lowest test score shown is in cell _____.

7. Circle the correct formula for calculating Beth's average score.

   D4 = (B4 + C4) / 2   D5 = B5 + C5   D5 = (B5 + C5) / 2

LESSON 3·8

How Far Can You See?

Suppose you are outdoors in a flat place where your view is not blocked by buildings or trees. You can see objects at ground level for several miles. The higher you are above ground or water level, the farther you can see. The distance you can see along Earth's surface is limited because Earth is curved. The **horizon**—where Earth and sky appear to intersect—is the farthest you can see along Earth's surface.

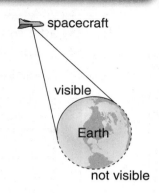

spacecraft

visible

Earth

not visible

The formula $d = 1.25 * \sqrt{h}$ gives the approximate distance d in miles you can see on a clear day, when h is the height of your eyes above ground or water level, measured in feet.

Reminder: $\sqrt{\ }$ means **square root** of. For example, 2 is the square root of 4 ($\sqrt{4} = 2$), because $2 * 2 = 4$.

1. You are standing on a boat deck. Your eyes are 15 feet above water level. As you look across the water, about how far can you see? (Round to the nearest mile.)

   The distance is about _____ miles.

2. Use the formula to fill in the table. Calculate the distance to the nearest tenth mile. Then add two more locations and calculate the distances. Use a reference book to find the heights of interesting places.

Place	Height	Distance		
Observation deck, Eiffel Tower, Paris, France	900 feet	About	_____	miles
Top of Sears Tower, Chicago, Illinois, U.S.A.	1,454 feet	About	_____	miles
Airplane in flight	30,000 feet	About	_____	miles
		About	_____	miles
		About	_____	miles

Try This

3. If you see the horizon about 10 miles away, your eye is about _____ above ground level.

(unit)

Math Boxes

1. The spreadsheet shows the number of baskets and free throws scored by players on a basketball team. Each basket is worth 2 points, and each free throw is worth 1 point. Complete the spreadsheet below.

	A	B	C	D
1	Player	Baskets	Free Throws	Total Points
2	Dion	1	2	4
3	Fran	5	0	
4	Sam	8	4	
5	**Total**			

a. What is shown in cell B3?

_____

b. Circle the formula for calculating the number of points Sam scored.

D4 = A4 + B4 + C4   D4 = (2 ∗ B4) + C4   D4 = B4 + (2 ∗ C4)

SRB
142–144

2. Divide.

20)‾365

365 ÷ 20 = _____

SRB
22–24

3. Use >, <, or = to compare each pair of numbers.

a. 0.0347 _____ 0.347

b. 76.203 _____ 76.2027

SRB
26 27

4. List the first 6 multiples of each number.

20 _____

25 _____

Name the least common multiple (LCM) of 20 and 25.

_____

SRB
78

5. List the factors of each number.

12 _____

18 _____

Name the greatest common factor (GCF) of 12 and 18.

_____

SRB
80

114

LESSON 3·9   # Matching Situations with Graphs

Math Message

Match each situation with the graph that represents it.

1. **Situation 1:** A car is parked in a driveway.   Graph _____

2. **Situation 2:** A car is traveling at a constant speed.   Graph _____

3. **Situation 3:** A car leaves a highway tollbooth.   Graph _____

4. **Situation 4:** A car approaches a red traffic light.   Graph _____

Graphs

Graph A

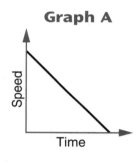

Graph B

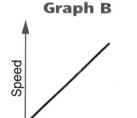

Graph C

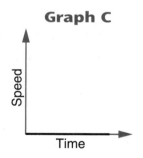

Graph D

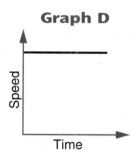

5. Draw a graph to represent the following situation.

A woman walks up one side of a hill at a steady pace and runs down the other side. She then continues walking at a steady pace.

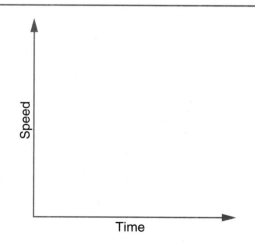

LESSON 3·9   Time Graphs

1. Mr. Olds drove his son Hank to school. The trip from home to school took 13 minutes. The graph below shows the speed that Mr. Olds traveled as he drove. Write a story that explains the shape of the graph.

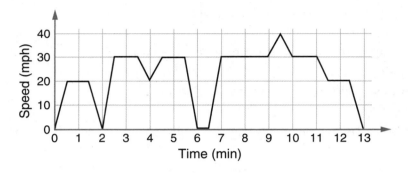

_____

_____

_____

_____

_____

2. Monica filled a cup with cocoa. She drank almost half of it. She refilled the cup. Then she drank all of the cocoa in the cup. Draw a graph that illustrates this story.

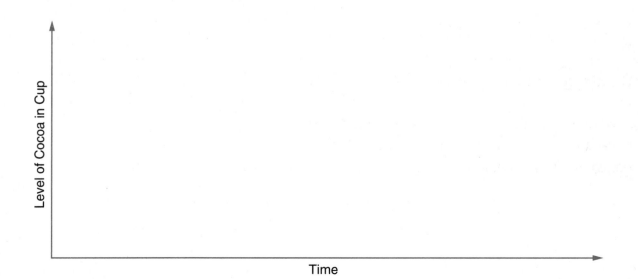

Mystery Graphs

Each of the graphs at the right represents one of the situations
described below. Match each situation with its graph.

Graph A

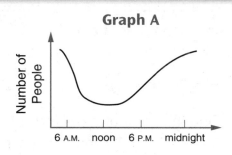

1. The number of people at a
school is best described by Graph _____.

Graph B

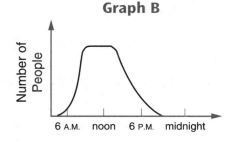

2. The number of people in a
restaurant is best described by Graph _____.

Graph C

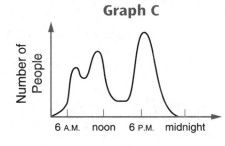

3. The number of people who are
at home is best described by Graph _____.

Graph D

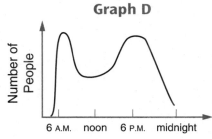

4. The number of people in a
hospital is best described by Graph _____.

5. The number of people driving
a car is best described by Graph _____.

Graph E

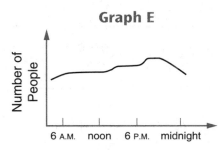

6. For one of the situations above, explain why you
chose to match that situation with a particular graph.

_____

_____

_____

Date _____ Time _____

---

1. The perimeter of a rectangle can be found by the formula

$P = 2 * (b + h).$

Find the perimeter (P) if $b = 5.5$ cm and $h = 3.25$ cm. Fill in the circle next to the best answer.

○ **A.** 3.77 cm   ○ **B.** 7.60 cm

○ **C.** 8.75 cm   ○ **D.** 17.50 cm

SRB
212

---

2. Give 3 special cases for the general pattern below.

$f + m^2 = f + (m * m)$

_____

_____

_____

SRB
103

---

3. Complete the table. Then graph the data and connect the plotted points.

Eddie travels about 8 miles per hour on his bike.

Rule: Distance traveled = 8 * number of hours

Time in hours (h)	Distance in miles ($8 * h$)
0	0
1	
2	
3	24
5	
	56

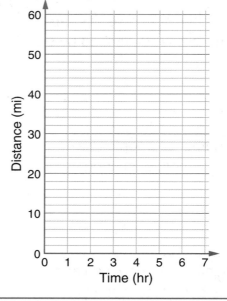

SRB
254

---

4. Rename each percent as a decimal.

a. 81% = _____

b. 7% = _____

c. 100% = _____

d. 70% = _____

SRB
59 60

---

5. List the first 6 multiples of each number.

8 _____

16 _____

24 _____

Name the least common multiple (LCM) of 8, 16, and 24.

_____

SRB
78

118

LESSON 3·10 Math Boxes

1. Darin charges $5 an hour to baby-sit on weekdays and $7 an hour on weekends. The spreadsheet is a record of the baby-sitting Darin did during one week. Complete the spreadsheet.

	A	B	C
1	Day of the Week	Number of Hours	Earnings ($)
2	Monday	4	
3	Wednesday	2	
4	Saturday	5	
5	**Total**		

a. Which cell contains the number of hours Darin worked on Saturday? _____

b. Circle the formula Darin should NOT use to calculate his total earnings.

C5 = C2 + C3 + C4   C5 = 6 * B5   C5 = (5 * B2) + (5 * B3) + (7 * B4)

c. Write a formula that Darin can use to calculate his earnings for Monday.

_____

SRB 142–144

2. Divide.

18)435.6

435.6 ÷ 18 = _____

SRB 42

3. Order from least to greatest.

a. 0.23, 0.32, 0.023, 0.323

_____

b. 36.837, 36.783, 36.878, 36.8375

_____

_____

SRB 30

4. List the first 6 multiples of each number.

5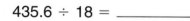

6

Name the least common multiple (LCM) of 5 and 6.

_____

SRB 78

5. Name the greatest common factor (GCF) of each pair of numbers.

a. 9 and 36 _____

b. 50 and 20 _____

c. 18 and 7 _____

SRB 80

LESSON 3·10 Formulas, Tables, and Graphs

Haylee and Chloe want to earn money during summer vacation.

Haylee's Summer Job	Chloe's Summer Job
Haylee is going to mow lawns. Her father will lend her $190 to buy a lawn mower. She figures that she can mow 10 lawns per week and make $12 per lawn after paying for oil and gasoline.	Chloe is going to work in an ice cream shop. The owner will provide a uniform free of charge and pay her $5.20 per hour. She will work $3\frac{1}{2}$ hours per day, 5 days per week.

1. Complete the table at the right to show how much profit each girl will have made after 2 weeks, 3 weeks, and so on. (Assume they do not have to pay taxes.)

2. Use the table to answer the following questions.

   a. Who will have made more money by the end of 3 weeks?

   _____

   b. How much money will that girl have made?

   _____

   c. Who will make more money during the summer?

   _____

Time (weeks)	Profit (dollars)	
	Haylee	Chloe
Start	−190	0
1	−70	91
2		
3		
4		
5		
6	530	
7		
8		728
9		
10		

LESSON 3·10 **Formulas, Tables, and Graphs** *continued*

3. Use the grid below to graph the profits from Haylee's and Chloe's summer jobs. Label the lines *Haylee* and *Chloe*.

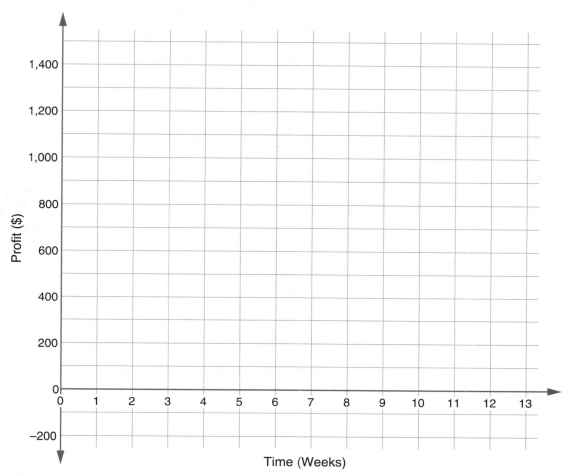

4. One way to analyze data is to look at how quickly the graph of the data rises or falls.

   a. Which graph rises more quickly, Haylee's graph or Chloe's graph?

   _____

   b. What is one conclusion that you can draw about profit data that are represented by a quickly rising graph?

   _____

   _____

   _____

5. Estimate the time at which the graphs intersect. _____

LESSON
3·11
Math Boxes

1. Compare each pair of fractions.
Write < or >.

a. $\frac{23}{25}$ _____ $\frac{21}{25}$

b. $\frac{3}{10}$ _____ $\frac{7}{20}$

c. $\frac{1}{7}$ _____ $\frac{1}{8}$

d. $\frac{5}{6}$ _____ $\frac{5}{9}$

 SRB 75

2. List all the factors of each number.

24 _____

36 _____

Name the greatest common factor (GCF)
of 24 and 36.

_____

 SRB 80

3. Rename each fraction as a decimal.

a. $\frac{1}{5}$ = _____

b. $\frac{4}{25}$ = _____

Rename each decimal as a fraction.

c. 0.02 = _____

d. 0.20 = _____

 SRB 59 60

4. List the first 6 multiples of each number.

12 _____

15 _____

Name the least common multiple (LCM)
of 12 and 15.

_____

 SRB 78

5. Rename each mixed number as a fraction.

a. $3\frac{2}{3}$ _____

b. $7\frac{3}{4}$ _____

Rename each fraction as a mixed or
whole number.

c. $\frac{9}{2}$ _____   d. $\frac{23}{8}$ _____

 SRB 71 72

6. Complete.

a. $\frac{2}{5} = \frac{\boxed{}}{50}$

b. $\frac{7}{\boxed{}} = \frac{21}{27}$

c. $\frac{\boxed{}}{8} = \frac{35}{56}$

d. $\frac{11}{16} = \frac{33}{\boxed{}}$

 SRB 73

Date _____   Time _____

Math Message

1. Fold a sheet of paper into 3 equal parts.
   Unfold it and shade 2 of the parts. What fraction of the paper is shaded? _____

2. Fold the paper back into thirds. Without unfolding it, fold it in half the other way.
   Then fold it in half again.

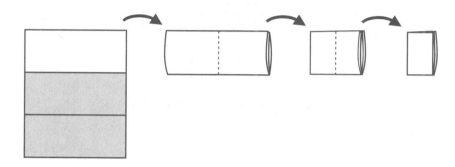

   a. How many rectangles are there? _____ rectangles

   b. How many are shaded? _____ rectangles

      $\frac{8}{12}$ of the paper is shaded. $\frac{2}{3}$ and $\frac{8}{12}$ are **equivalent fractions**.

3. Use paper folding to find a fraction equivalent to $\frac{3}{4}$.

   a. Fold a sheet of paper into 4 equal parts. Unfold it and
      shade 3 of the parts. What fraction of the paper is shaded? _____

   b. Draw a picture of the sheet of paper
      that shows $\frac{3}{4}$ shaded.

   c. Fold the paper to create an
      equivalent fraction.

   d. Draw a picture of the paper that shows
      the equivalent fraction you made.

   e. How many equal parts are there? _____ equal parts

   f. How many parts are shaded? _____ parts

   g. Name the equivalent fraction. _____

123

LESSON 4·1 **Equivalent Fractions**

You can find equivalent fractions in the following ways:

Multiplication Rule	Division Rule
To find an equivalent fraction, *multiply* the numerator and the denominator of the original fraction by the same (nonzero) number.	To find an equivalent fraction, *divide* the numerator and the denominator of the original fraction by the same (nonzero) number.
Example: $\frac{2}{3}$ $\frac{2 * 2}{3 * 2} = \frac{4}{6}$	Example: $\frac{20}{12}$ $\frac{20 \div 4}{12 \div 4} = \frac{5}{3}$

1. Find an equivalent fraction by multiplying.

   a. $\frac{3}{5}$ _____

   b. $\frac{5}{3}$ _____

   c. $\frac{3}{10}$ _____

   d. $\frac{3}{4}$ _____

   e. $\frac{2}{5}$ _____

2. Find an equivalent fraction by dividing.

   a. $\frac{6}{8}$ _____

   b. $\frac{8}{12}$ _____

   c. $\frac{50}{75}$ _____

   d. $\frac{30}{100}$ _____

   e. $\frac{36}{24}$ _____

3. Write three equivalent fractions for each given fraction.

   a. $\frac{2}{3}$ _____

   b. $\frac{5}{10}$ _____

   c. $\frac{5}{8}$ _____

   d. $\frac{50}{45}$ _____

   e. $\frac{45}{50}$ _____

   f. $\frac{16}{28}$ _____

4. Any whole number can be written as a fraction. For example: $2 = \frac{4}{2} = \frac{6}{3} = \frac{8}{4}$.
   Write at least two equivalent fractions for each whole number.

   a. 3 _____

   b. 4 _____

   c. 0 _____

   d. 1 _____

LESSON 4·1 Equivalent Fractions *continued*

A fraction is in **simplest form** if its numerator and denominator do not have any factors in common except 1.

For example, $\frac{2}{3}$ is in simplest form because the only **common factor** of 2 and 3 is 1.
$\frac{4}{8}$ is not in simplest form, because 4 is a common factor of 4 and 8.

5. Circle the fractions that are in simplest form.

a. $\frac{3}{5}$ b. $\frac{4}{6}$ c. $\frac{9}{14}$ d. $\frac{12}{13}$ e. $\frac{12}{15}$ f. $\frac{8}{18}$

6. One way to find the simplest form of a fraction is to divide its numerator and denominator by the **greatest common factor** of the 2 numbers.

Example 1:
Rename $\frac{8}{12}$ as a fraction in simplest form.
The greatest common factor of 8 and 12 is 4. $\frac{8}{12} = \frac{8 \div 4}{12 \div 4} = \frac{2}{3}$

Write each fraction in simplest form.

a. $\frac{12}{16}$ = _____   b. $\frac{3}{9}$ = _____   c. $\frac{8}{14}$ = _____   d. $\frac{9}{15}$ = _____

e. $\frac{18}{45}$ = _____   f. $\frac{24}{32}$ = _____   g. $\frac{9}{24}$ = _____   h. $\frac{20}{25}$ = _____

7. Find the missing numbers.

Example 2: $\frac{3}{4} = \frac{x}{8}$

The denominator of the second fraction is twice as much as the denominator of the first fraction. Therefore, the numerator of the second fraction must also be twice as much as the numerator of the first fraction.

$\frac{3}{4} = \frac{3 * 2}{4 * 2} = \frac{6}{8}$; $x = 6$

Example 3: $\frac{9}{15} = \frac{3}{y}$

The numerator of the second fraction is one-third as much as the numerator of the first fraction. Therefore, the denominator of the second fraction must also be one-third as much as the denominator of the first fraction.

$\frac{9}{15} = \frac{9 \div 3}{15 \div 3} = \frac{3}{5}$; $y = 5$

a. $\frac{1}{4} = \frac{3}{y}$ $y =$ _____   b. $\frac{3}{z} = \frac{30}{50}$ $z =$ _____   c. $\frac{w}{8} = \frac{24}{32}$ $w =$ _____

d. $\frac{2}{3} = \frac{10}{x}$ $x =$ _____   e. $\frac{d}{4} = \frac{18}{24}$ $d =$ _____   f. $\frac{15}{25} = \frac{r}{5}$ $r =$ _____

LESSON 4·1   Math Boxes

---

1. Name the greatest common factor for each pair of numbers.

a. 2 and 5 _____

b. 9 and 18 _____

c. 20 and 32 _____

d. 54 and 63 _____

SRB 80

2. Find the missing numbers.

a. $\dfrac{2}{1} = \dfrac{d}{25}$ $d = $ _____

b. $\dfrac{9}{m} = \dfrac{81}{18}$ $m = $ _____

c. $\dfrac{16}{52} = \dfrac{t}{13}$ $t = $ _____

d. $\dfrac{7}{10} = \dfrac{28}{k}$ $k = $ _____

SRB 73

---

3. The table shows the ages of members in a book club. Make a stem-and-leaf plot to display the members' ages.

Ages of Book Club Members						
45	27	38	34	22	46	50
29	53	24	41	64	43	36

Use your stem-and-leaf plot to find the median age.   median _____

Ages of Book Club Members	
Stems (10s)	**Leaves** (1s)

SRB 135

---

4. Add.

a. $24 + (-8) = $ _____

b. $19 + (-26) = $ _____

c. $-34 + (-18) = $ _____

d. $-21 + 12 = $ _____

SRB 95

5. Tell whether each angle is acute, right, obtuse, reflex, or straight.

a. 120° angle _____

b. 15° angle _____

c. 90° angle _____

d. 180° angle _____

SRB 160

LESSON 4·2 Comparing Fractions

Math Message

You can make any fraction using the digits 0, 1, 2, 3, 4, 5, 6, 7, 8, and 9.
The fraction $\frac{3}{4}$ is made up of 2 digits; the fraction $\frac{23}{6}$ is made up of 3 digits.
A fraction never has a denominator of 0.

Use only 2 digits to make the following fractions.

1. The least possible fraction that is greater than 0 _____

2. The greatest possible fraction _____

3. The greatest possible fraction that is less than 1 _____

4. The least possible fraction that is greater than $\frac{1}{2}$ _____

Compare each pair of fractions by first renaming them with a common denominator.

Write $<$, $>$, or $=$. Show how you got each answer.

Example: $\frac{3}{4}$ _____ $\frac{5}{6}$

One way: The QCD is $4 * 6 = 24$. Rename each fraction so the denominator is 24.	Another way: The LCD of 4 and 6 is 12. Rename each fraction so the denominator is 12.
$\frac{3}{4} = \frac{x}{24}$; $x = 18$ $\frac{5}{6} = \frac{y}{24}$; $y = 20$	$\frac{3}{4} = \frac{x}{12}$; $x = 9$ $\frac{5}{6} = \frac{x}{12}$; $x = 10$
$\frac{3}{4} = \frac{18}{24}$ $\frac{5}{6} = \frac{20}{24}$	$\frac{3}{4} = \frac{9}{12}$ $\frac{5}{6} = \frac{10}{12}$
$\frac{18}{24} < \frac{20}{24}$, so $\frac{3}{4} < \frac{5}{6}$	$\frac{9}{12} < \frac{10}{12}$, so $\frac{3}{4} < \frac{5}{6}$

5. $\frac{2}{3}$ _____ $\frac{4}{5}$

6. $\frac{2}{5}$ _____ $\frac{3}{10}$

7. $\frac{1}{3}$ _____ $\frac{5}{8}$

8. $\frac{2}{6}$ _____ $\frac{1}{5}$

9. $\frac{3}{4}$ _____ $\frac{18}{24}$

10. $\frac{7}{6}$ _____ $\frac{11}{9}$

LESSON 4·2 Math Boxes

1. Write each fraction in simplest form.

  a. $\frac{6}{8}$ _____   **b.** $\frac{21}{36}$ _____

  c. $\frac{45}{50}$ _____   **d.** $\frac{27}{51}$ _____

  Write 2 fractions equivalent to $\frac{3}{5}$.

  e. _____   _____

2. Add or subtract. Then simplify.

  a. $\frac{1}{5} + \frac{4}{5} =$ _____

  b. $\frac{3}{6} + \frac{1}{2} =$ _____

  c. $\frac{7}{9} - \frac{4}{9} =$ _____

  d. $\frac{6}{8} - \frac{3}{4} =$ _____

3. Find the median and mean for the following set of numbers.

  123, 56, 92, 90, 88, 91

  a. median _____   **b.** mean _____

  Suppose the minimum of the data set was 60 instead of 56. Which landmark, the median or the mean, would change?

  _____

4. Write an algebraic expression for each phrase.

  a. 5 less than a number x _____

  b. one-eighth of a number m

  _____

  c. 9 more than a number p _____

5. The table below shows the results of a survey in which sixth graders were asked what they did when they caught a cold. Use a Percent Circle to make a circle graph of the results.

Ways of Dealing with a Cold

Ways of Dealing with a Cold	Percent of Sixth Graders Surveyed
Do nothing	62%
Rest and fluids	17%
Medicine	13%
Other	8%

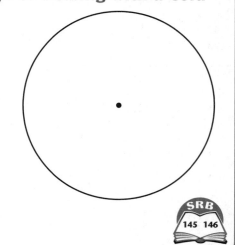

Date _____   Time _____

Fractions of a Square

Math Message

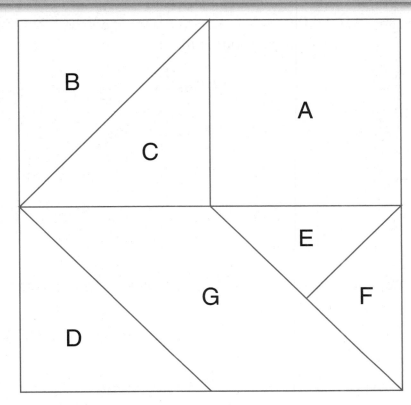

1. What fraction of the large square is …

   a. Square A? _____

   b. Triangle B? _____

   c. Triangle E? _____

   d. Parallelogram G? _____

2. What fraction of the large square are the following pieces, when put together?
   Write a number sentence to show your answer.

   a. Triangles B and C _____

   b. Triangles E and F _____

   c. Square A and Triangle C _____

   d. Square A and Triangle E _____

   e. Triangles E and B _____

   f. Square A and Parallelogram G _____

   g. Triangles D, E, and F and Parallelogram G

   _____

129

LESSON 4·3 Adding and Subtracting Fractions

To add or subtract fractions with different denominators, first find equivalent fractions with a common denominator. Then add or subtract the numerators.

Example 1: $\frac{1}{8} + \frac{3}{8} = \frac{1+3}{8} = \frac{4}{8} = \frac{1}{2}$ **Example 2:** $\frac{2}{3} - \frac{1}{4} = \frac{8}{12} - \frac{3}{12} = \frac{8-3}{12} = \frac{5}{12}$

Add or subtract. Write your answers as fractions in simplest form.

1. $\frac{1}{2} + \frac{1}{4} =$ _____

2. $\frac{1}{3} + \frac{1}{2} =$ _____

3. $1 - \frac{1}{4} =$ _____

4. $\frac{1}{2} + \frac{1}{6} =$ _____

5. $\frac{3}{4} - \frac{1}{6} =$ _____

6. $\frac{5}{6} + \frac{1}{12} =$ _____

7. $\frac{1}{8} + \frac{1}{4} =$ _____

8. $\frac{2}{3} + \frac{5}{6} =$ _____

9. $\frac{1}{3} + \frac{1}{6} - \frac{2}{12} =$ _____

For each of the following, estimate whether the sum is greater than 1 or less than 1. Then calculate the sum and check your estimate.

	> 1 or < 1	Sum		> 1 or < 1	Sum
10. $\frac{3}{4} + \frac{3}{8}$	_____	_____	**11.** $\frac{1}{2} + \frac{2}{3}$	_____	_____
12. $\frac{3}{5} + \frac{1}{10}$	_____	_____	**13.** $\frac{1}{8} + \frac{1}{5}$	_____	_____

14. Mentally estimate the sums and differences below. Write the letter of the problem by the point on the number line closest to your estimate. Study **14a** as an example.

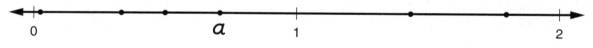

a. $\frac{3}{8} + \frac{1}{3}$

b. $\frac{7}{8} + \frac{12}{13}$

c. $\frac{9}{10} - \frac{7}{8}$

d. $\frac{1}{3} + \frac{1}{6}$

e. $\frac{5}{6} + \frac{3}{5}$

f. $\frac{7}{12} - \frac{1}{4}$

15. Name two fractions that have a sum

a. between 1 and 2.

b. between $\frac{1}{2}$ and 1.

c. between $\frac{1}{2}$ and $\frac{3}{4}$.

_____ _____ _____

LESSON 4·3 **Math Boxes**

---

1. Name the greatest common factor for each pair of numbers.

a. 12 and 30 _____

b. 18 and 45 _____

c. 27 and 51 _____

d. 17 and 68 _____

SRB 80

---

2. Find the missing numbers.

a. $\frac{12}{20} = \frac{108}{y}$ $y =$ _____

b. $\frac{31}{b} = \frac{124}{208}$ $b =$ _____

c. $\frac{0.4}{2.4} = \frac{f}{6}$ $f =$ _____

d. $\frac{m}{360} = \frac{108}{36}$ $m =$ _____

SRB 73

---

3. The table shows the heights in centimeters of Mr. Carmen's dance students. Make a stem-and-leaf plot to display the students' heights.

Students' Heights (cm)				
173	158	188	165	170
168	163	168	173	183

Heights of Dance Students

Stems (100s and 10s)	Leaves (1s)

Use your stem-and-leaf plot to find the median height.   median _____

SRB 135

---

4. Add.

a. $46 + (-13) =$ _____

b. $25 + (-76) =$ _____

c. $-27 + (-44) =$ _____

d. $-46 + 37 =$ _____

SRB 95

---

5. Which of the angles is a reflex angle? Fill in the circle next to the best answer.

Ⓐ 270° angle

Ⓑ 10° angle

Ⓒ 130° angle

Ⓓ $40\frac{1}{2}$° angle

SRB 160

LESSON 4·4

Adding and Subtracting Mixed Numbers

SRB
84–86

Example 1: $1\frac{2}{5} + 2\frac{4}{5} = ?$

Step 1	**Step 2**
Add the fractions. Then add the whole numbers. $$\begin{array}{r} 1\frac{2}{5} \\ + 2\frac{4}{5} \\ \hline 3\frac{6}{5} \end{array}$$	If necessary, rename the sum. $$\begin{aligned} 3\frac{6}{5} &= 3 + \frac{6}{5} \\ &= 3 + \frac{5}{5} + \frac{1}{5} \\ &= 3 + 1 + \frac{1}{5} \\ &= 4\frac{1}{5} \end{aligned}$$

Add. Write your answers in simplest form.

1. $$\begin{array}{r} 4\frac{1}{5} \\ + 3\frac{2}{5} \\ \hline \end{array}$$
2. $$\begin{array}{r} 1\frac{2}{4} \\ + 2\frac{3}{4} \\ \hline \end{array}$$
3. $$\begin{array}{r} 5\frac{1}{4} \\ + 1\frac{3}{4} \\ \hline \end{array}$$
4. $$\begin{array}{r} 1\frac{4}{8} \\ + 1\frac{2}{8} \\ \hline \end{array}$$

Example 2: $4\frac{5}{8} - 2\frac{1}{8} = ?$

Step 1	**Step 2**
Subtract the fractions. Then subtract the whole numbers. $$\begin{array}{r} 4\frac{5}{8} \\ - 2\frac{1}{8} \\ \hline 2\frac{4}{8} \end{array}$$	If necessary, rename the difference. $$\begin{array}{r} 4\frac{5}{8} \\ - 2\frac{1}{8} \\ \hline 2\frac{4}{8} = 2\frac{1}{2} \end{array}$$

Example 3: $5\frac{1}{3} - 1\frac{2}{3} = ?$

Notice that the fraction in the first mixed number is less than the fraction in the second mixed number. Because you can't subtract $\frac{2}{3}$ from $\frac{1}{3}$, you need to rename $5\frac{1}{3}$.

Step 1	**Step 2**
Rename the first mixed number. $$\begin{aligned} 5\frac{1}{3} &= 4 + 1 + \frac{1}{3} \\ &= 4 + \frac{3}{3} + \frac{1}{3} \\ &= 4 + \frac{4}{3} = 4\frac{4}{3} \end{aligned}$$	Subtract the fractions. Then subtract the whole numbers. $$\begin{array}{r} 5\frac{1}{3} \longrightarrow 4\frac{4}{3} \\ - 1\frac{2}{3} \qquad - 1\frac{2}{3} \\ \hline 3\frac{2}{3} \end{array}$$

LESSON 4·4   **Adding and Subtracting Mixed Numbers** *cont.*

SRB 84–86

Example 4: $8 - 3\frac{5}{8} = ?$

Step 1	**Step 2**
Rename the whole number.	Subtract the fractions. Then subtract the whole numbers.
$8 = 7 + 1$	$\begin{array}{r} 8 \longrightarrow 7\frac{8}{8} \\ -\ 3\frac{5}{8} \quad -\ 3\frac{5}{8} \\ \hline 4\frac{3}{8} \end{array}$
$= 7 + \frac{8}{8}$	
$= 7\frac{8}{8}$	

Add or subtract.

5. $\begin{array}{r} 3\frac{1}{4} \\ -\ 2\frac{3}{4} \\ \hline \end{array}$

6. $\begin{array}{r} 4\frac{1}{5} \\ -\ 2\frac{2}{5} \\ \hline \end{array}$

7. $\begin{array}{r} 5 \\ -\ 2\frac{2}{3} \\ \hline \end{array}$

8. $\begin{array}{r} 7 \\ -\ 3\frac{1}{4} \\ \hline \end{array}$

9. $\begin{array}{r} 4\frac{1}{6} \\ +\ 2\frac{1}{6} \\ \hline \end{array}$

10. $\begin{array}{r} 3\frac{1}{5} \\ -\ 1\frac{3}{5} \\ \hline \end{array}$

11. $\begin{array}{r} 5 \\ -\ 1\frac{3}{4} \\ \hline \end{array}$

12. $\begin{array}{r} 6\frac{7}{8} \\ +\ 3\frac{3}{8} \\ \hline \end{array}$

13. Joe has a board that is $8\frac{3}{4}$ inches long. He cuts off $1\frac{1}{4}$ inches. How long is the remaining piece? _____

14. Mr. Ventrelli is making bread. He adds $1\frac{1}{4}$ cups of white flour and $1\frac{1}{4}$ cups of wheat flour. The recipe calls for the same number of cups of water as cups of flour. How much water should he add? _____

15. Evelyn's house is between Robert's and Elizabeth's. How far is Robert's house from Elizabeth's?

_____

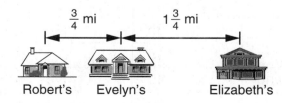

Robert's   Evelyn's   Elizabeth's

LESSON 4·4   **Math Boxes**

---

1. Write each fraction in simplest form.

   a. $\frac{18}{45}$ = _____   b. $\frac{26}{39}$ = _____

   c. $\frac{56}{80}$ = _____   d. $\frac{25}{625}$ = _____

   Write 2 fractions equivalent to $\frac{9}{4}$.

   e. _____   _____

   74

2. Add or subtract. Then simplify.

   a. $\frac{1}{10} + \frac{3}{5}$ = _____

   b. $\frac{5}{12} + \frac{1}{3}$ = _____

   c. $\frac{7}{9} - \frac{4}{9}$ = _____

   d. $\frac{6}{8} - \frac{3}{4}$ = _____

   83

---

3. Find the median and mean for the following set of numbers.

   1.5, 2.8, 3.4, 4.5, 2.2, 8.4

   a. median _____   b. mean _____

   Suppose you multiplied each data value by 2. What would happen to the mean?

   _____

   136 137

4. Thomas Jefferson was born in 1743. George Washington was born m years earlier. In what year was Washington born? Choose the best answer.

   ⬭ $m + 1743$

   ⬭ $m - 1743$

   ⬭ $1743 - m$

   ⬭ $1743 + m$

   240

---

5. The table below shows the results of a survey in which people were asked which winter Olympic sport they most enjoyed watching. Use a Percent Circle to make a circle graph of the results.

Favorite Sport	Percent of People Surveyed
Luge	35%
Ice hockey	15%
Figure skating	40%
Other	10%

Winter Olympic Sports Preferences

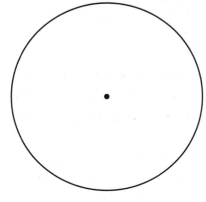

145 146

LESSON 4·5   **Math Boxes**

---

1. Rename each mixed number in simplest form.

   a. $1\frac{3}{3}$ = _____

   b. _____ = $2\frac{13}{6}$

   c. $3\frac{28}{5}$ = _____

   d. _____ = $6\frac{43}{3}$

   SRB
   72

2. Subtract. Write the answer as a mixed number in simplest form.

   a. $8\frac{7}{12} - \frac{10}{12}$ = _____

   b. $8\frac{9}{10} - 7\frac{3}{10}$ = _____

   c. $5\frac{7}{16} - 3\frac{3}{8}$ = _____

   d. $6\frac{2}{3} - 2\frac{1}{4}$ = _____

   SRB
   85 86

---

3. Multiply and divide mentally.

   a. $0.8 * 100$ = _____

   b. $5.03 \div 10$ = _____

   c. _____ = $27.5 \div 100$

   d. _____ = $0.0918 * 100{,}000$

   SRB
   35 36
   40 41

4. Write the products in standard notation.

   a. _____ = $3.67 * 10^4$

   b. $2.01 * 10^{-1}$ = _____

   c. $45.2 * 10^{-2}$ = _____

   d. _____ = $0.0443 * 10^5$

   SRB
   8

---

5. Divide.

   $12\overline{)303.6}$

   $303.6 \div 12$ = _____

   SRB
   42–45

6. Draw and label the following angle.

   $\angle POL = 135°$

   SRB
   230 232

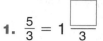

More Adding and Subtracting Mixed Numbers

Fill in the missing number to rename the mixed number.

1. $\frac{5}{3} = 1\frac{\boxed{}}{3}$

2. $3\frac{7}{4} = 4\frac{\boxed{}}{4}$

3. $6\frac{12}{8} = \underline{}\frac{4}{8}$

4. $5\frac{2}{5} = 4\frac{\boxed{}}{5}$

5. $8\frac{2}{3} = 7\frac{\boxed{}}{3}$

6. $2\frac{3}{8} = 1\frac{\boxed{}}{8}$

Use what you know about adding and subtracting fractions to help you add and subtract mixed numbers with different denominators.

Example 1: $1\frac{1}{3} + 3\frac{5}{6} = ?$

Estimate: $1 + 4 = 5$

Rename $\frac{1}{3}$ as $\frac{2}{6}$.
Add and then simplify.

$$1\frac{1}{3} \longrightarrow 1\frac{2}{6}$$
$$\underline{+\,3\frac{5}{6}} \qquad \underline{+\,3\frac{5}{6}}$$
$$\phantom{+\,3\frac{5}{6}} \qquad 4\frac{7}{6} \longrightarrow 5\frac{1}{6}$$

Example 2: $8\frac{2}{5} - 2\frac{1}{2} = ?$

Estimate: $8\frac{1}{2} - 2\frac{1}{2} = 6$

Rename $\frac{2}{5}$ as $\frac{4}{10}$ and $\frac{1}{2}$ as $\frac{5}{10}$.
Rename $8\frac{4}{10}$ as $7\frac{14}{10}$. Then subtract.

$$8\frac{2}{5} \longrightarrow 8\frac{4}{10} \longrightarrow 7\frac{14}{10}$$
$$\underline{-\,2\frac{1}{2}} \qquad \underline{-\,2\frac{5}{10}} \qquad \underline{-\,2\frac{5}{10}}$$
$$\phantom{-\,2\frac{1}{2}} \qquad \phantom{-\,2\frac{5}{10}} \qquad 5\frac{9}{10}$$

First estimate. Then add or subtract. Show your work.

7.
$$1\frac{3}{4}$$
$$\underline{+\,3\frac{2}{8}}$$
Estimate: _____

8.
$$3\frac{1}{2}$$
$$\underline{-\,1\frac{1}{4}}$$
Estimate: _____

9.
$$6$$
$$\underline{-\,2\frac{1}{5}}$$
Estimate: _____

10.
$$2\frac{7}{8}$$
$$\underline{+\,1\frac{3}{4}}$$
Estimate: _____

11. $6\frac{1}{8} - 5\frac{1}{4} = ?$ Estimate: _____

$6\frac{1}{8} - 5\frac{1}{4} = $ _____

12. $5\frac{1}{3} - 4\frac{3}{4} = ?$ Estimate: _____

$5\frac{1}{3} - 4\frac{3}{4} = $ _____

LESSON 4·5   **More Adding and Subtracting Mixed Numbers** *cont.*

13. To check your answer to Problem 10, draw a line segment that is $2\frac{7}{8}$ inches long. Then make it $1\frac{3}{4}$ inches longer.

Measure the whole line segment. How long is it? _____

14. Tim walked $\frac{1}{2}$ mile from school to the park. Then he walked $\frac{1}{4}$ mile to the store to get milk and orange juice. These items cost $2\frac{1}{2}$ dollars. Then he walked another $\frac{3}{4}$ mile to go home.

   How far did Tim walk in all? _____

Find a way to solve these problems without finding common denominators.
Write number sentences to show the order in which you added and subtracted.

15. $1\frac{1}{2} + 3\frac{7}{8} + 4\frac{1}{2} + 1\frac{1}{8} =$ _____

_____

16. $3\frac{1}{2} + 4\frac{2}{3} - 1\frac{1}{2} =$ _____

_____

17. If you drive $5\frac{4}{5}$ miles from Alphatown along Highway 1, you reach Betaville. Driving farther, you reach Gamma, which is $8\frac{1}{10}$ miles from Alphatown.

   a. Plot the locations of Betaville and Gamma on the number line below.

   Alphatown

   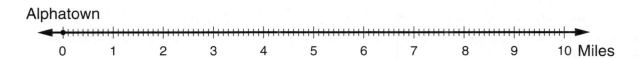

   b. How far is it from Betaville to Gamma? _____

LESSON 4·5 Water Problems

1. The largest ocean is the Pacific Ocean. Its area is about 64 million square miles. The second largest ocean is the Atlantic Ocean. Let P stand for the area of the Pacific Ocean. If $P - 30.6$ stands for the area of the Atlantic, then what is the approximate area of the Atlantic?

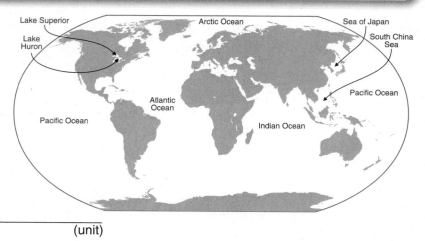

_____
(unit)

2. The third largest ocean is the Indian Ocean. Its area is about 28.3 million square miles. Write an algebraic expression using P that represents the approximate area of the Indian Ocean.

_____

3. The fourth largest ocean is the Arctic Ocean. If $\frac{P}{12}$ stands for the approximate area of the Arctic Ocean, then what is the approximate area of the Arctic Ocean?

_____
(unit)

4. Let A stand for the area of the Arctic Ocean. The area of the South China Sea, the largest sea in the world, is about $\frac{1}{5}$ the area of the Arctic Ocean.

   a. Use the variable A to write an algebraic expression that represents the area of the South China Sea.   _____

   b. What is the approximate area of the South China Sea?

_____
(unit)

5. The deepest point in the Sea of Japan is about 5,468 feet below sea level. The deepest point in the Indian Ocean is 7,534 feet deeper. About how many feet below sea level is the deepest point in the Indian Ocean?

_____
(unit)

6. The area of Lake Superior, the largest of the Great Lakes, is about 31,699 square miles. The area of Lake Huron, the second largest, is approximately 23,004 square miles. About how much larger is Lake Superior than Lake Huron?

_____
(unit)

LESSON 4·6   ## Math Boxes

---

1. Compare. Write $<$ or $>$.

a. $\dfrac{4}{9}$ _____ $\dfrac{3}{5}$

b. $\dfrac{3}{4}$ _____ $\dfrac{3}{8}$

c. $\dfrac{6}{7}$ _____ $\dfrac{4}{5}$

d. $\dfrac{1}{9}$ _____ $\dfrac{1}{10}$

SRB 75

2. Multiply. Write each answer in simplest form.

a. $\dfrac{1}{4} * \dfrac{3}{5} =$ _____

b. $\dfrac{1}{2} * \dfrac{7}{8} =$ _____

c. _____ $= \dfrac{3}{4} * \dfrac{5}{9}$

d. _____ $= \dfrac{2}{5} * \dfrac{4}{5}$

SRB 74 89

---

3. This spreadsheet shows the number of hours 3 students slept on 2 different nights.

a. Calculate the mean number of sleeping hours for each student. Write these means in Column D.

	A	B	C	D
1	Student	Monday	Tuesday	Mean
2	Allie	6.5	9	
3	Franco	6	10.5	
4	Blake	8	7	

b. Using cell names, write a formula for calculating Blake's mean number of sleeping hours. _____

SRB 142–144

---

4. Evaluate the following algebraic expressions for $b = \dfrac{1}{6}$.

a. $5\dfrac{2}{3} + b$ _____

b. $5 - b$ _____

c. $\dfrac{1}{2} - b$ _____

d. $8 * b$ _____

SRB 83 84 89

5. Measure the angle to the nearest degree.

$m\angle MOP$ is _____ °

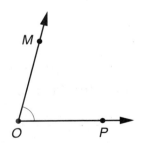

SRB 230 231

LESSON 4·6   # A Fraction Multiplication Algorithm

Math Message

1. Use the number line to help you solve the problems in Columns 1 and 2 below.
Reminder: The word *of* often means *times.*

```
  0                    1                    2                    3
```

Column 1

$\frac{1}{2}$ of 3 = _____

$\frac{1}{4}$ of $\frac{1}{2}$ = _____

$\frac{5}{8}$ of 1 = _____

$\frac{1}{3}$ of $\frac{3}{4}$ = _____

Column 2

$\frac{1}{2} * 3$ = _____

$\frac{1}{4} * \frac{1}{2}$ = _____

$\frac{5}{8} * 1$ = _____

$\frac{1}{3} * \frac{3}{4}$ = _____

2. How are the problems in Column 1 like their partner problems in Column 2?

_____

3. Circle the general pattern(s) below for the partner problems in Columns 1 and 2.

$a + b = b + a$ $a * b = b * c$ a of $b = a * b$ $a \div b = b \div a$

Six special cases of a general pattern are given below. Write the answer in simplest form
for each special case. Study the first one.

4. $\frac{1}{5} * \frac{2}{3} = \frac{1 * 2}{5 * 3} = \frac{2}{15}$

5. $\frac{3}{4} * \frac{1}{2} = \frac{3 * 1}{4 * 2} = $ _____

6. $\frac{2}{1} * \frac{2}{4} = \frac{2 * 2}{1 * 4} = $ _____

7. $\frac{2}{4} * \frac{3}{5} = \frac{2 * 3}{4 * 5} = $ _____

8. $\frac{4}{6} * \frac{1}{2} = \frac{4 * 1}{6 * 2} = $ _____

9. $\frac{3}{7} * \frac{1}{3} = \frac{3 * 1}{7 * 3} = $ _____

10. Describe the general pattern in words. (*Hint:* Look at the numerators and
denominators of the factors and products.)

_____

_____

_____

LESSON 4·6 **A Fraction Multiplication Algorithm** *cont.*

11. Try to write the general pattern for Problem 10 using variables.
(*Hint:* Use four variables.)

_____

Use the general pattern you found in Problem 11 to solve the following multiplication problems. Study the first one.

12. $\frac{3}{8} * \frac{2}{3} =$ _____ $\frac{3 * 2}{8 * 3} = \frac{6}{24}$ _____

13. $\frac{1}{3} * \frac{2}{3} =$ _____

14. $\frac{4}{5} * \frac{2}{8} =$ _____

15. $\frac{3}{12} * \frac{2}{4} =$ _____

16. $\frac{3}{4} * \frac{5}{6} =$ _____

17. $\frac{7}{9} * \frac{3}{8} =$ _____

18. $\frac{2}{5} * \frac{7}{8} =$ _____

19. $\frac{5}{10} * \frac{4}{7} =$ _____

Write the following whole numbers as fractions. The first one has been done for you.

20. $6 =$ _____ $\frac{6}{1}$ _____

21. $3 =$ _____

22. $5 =$ _____

23. $7 =$ _____

24. Rewrite the following problems as fraction multiplication problems and then solve them. Study the first one.

a. $4 * \frac{2}{3} =$ _____ $\frac{4}{1} * \frac{2}{3}$ _____ $= \frac{8}{3}$ _____

b. $6 * \frac{3}{5} =$ _____ $=$ _____

c. $7 * \frac{5}{6} =$ _____ $=$ _____

d. $3 * \frac{3}{4} =$ _____ $=$ _____

Try This

25. Write a general pattern with variables for the special cases in Problem 24.
(*Hint:* Use three variables.)

_____

26. Mark took a timed multiplication test and finished $\frac{3}{4}$ of the problems.
He correctly answered $\frac{1}{2}$ of the problems he finished. What fraction
of the problems on the test did Mark answer correctly?

_____

LESSON 4·6 **Mixed-Number Addition**

1. Rename each mixed number as a fraction.

   a. $2\frac{3}{4}$ _____

   b. $5\frac{6}{10}$ _____

   c. $4\frac{1}{3}$ _____

   d. $2\frac{4}{7}$ _____

   e. $8\frac{3}{5}$ _____

2. Rename each fraction as a mixed number or whole number.

   a. $\frac{17}{3}$ _____

   b. $\frac{14}{7}$ _____

   c. $\frac{16}{5}$ _____

   d. $\frac{41}{8}$ _____

   e. $\frac{58}{6}$ _____

Add. Write each answer as a whole number or a mixed number in simplest form.

3. $5\frac{3}{8} + 7\frac{7}{8} =$ _____

4. $3\frac{1}{2} + 2\frac{4}{5} =$ _____

5. $1\frac{1}{6} + 3\frac{3}{4} =$ _____

6. $9\frac{1}{5} + 3\frac{1}{2} =$ _____

7. $2\frac{2}{5} + 4\frac{1}{3} =$ _____

8. $1\frac{6}{10} + 2\frac{2}{5} =$ _____

9. $3\frac{2}{7} + 5 =$ _____

10. $4\frac{9}{10} + \frac{7}{20} =$ _____

Date _____ Time _____

LESSON 4·6   **Mixed-Number Subtraction**

1. Rename each mixed number as a fraction.

   a. $2\frac{5}{6}$ _____

   b. $10\frac{4}{5}$ _____

   c. $4\frac{1}{2}$ _____

   d. $3\frac{1}{6}$ _____

   e. $8\frac{3}{4}$ _____

2. Rename each fraction as a mixed number or whole number.

   a. $\frac{15}{5}$ _____

   b. $\frac{7}{7}$ _____

   c. $\frac{98}{5}$ _____

   d. $\frac{45}{10}$ _____

   e. $\frac{3}{2}$ _____

Subtract. If possible, write answers as mixed or whole numbers in simplest form.

3. $1\frac{1}{3} - \frac{2}{3} =$ _____

4. $4\frac{3}{4} - 2\frac{1}{4} =$ _____

5. $4 - 2\frac{1}{5} =$ _____

6. $3\frac{5}{6} - 1\frac{1}{3} =$ _____

7. $2\frac{1}{5} - \frac{4}{5} =$ _____

8. $15\frac{5}{7} - 13\frac{4}{7} =$ _____

9. $5 - 3\frac{2}{3} =$ _____

10. $10\frac{3}{5} - 6\frac{3}{10} =$ _____

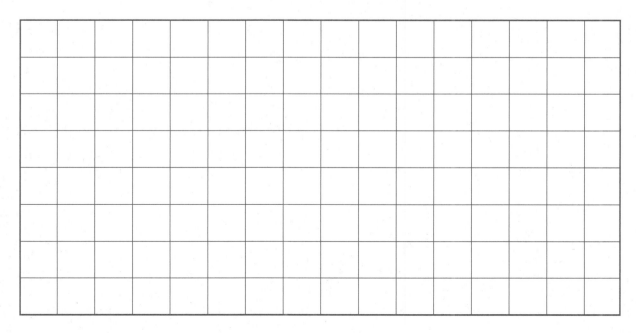

143

LESSON
4·7

Multiplying Mixed Numbers

Here are two methods for multiplying mixed numbers.

Example: $3\frac{3}{4} * 2\frac{2}{3} = ?$

Partial-Products Method	**Renaming-Mixed-Numbers Method**
$3\frac{3}{4}$ $* 2\frac{2}{3}$ $2 * 3 = 6$ $2 * \frac{3}{4} = \frac{6}{4} = 1 + \frac{1}{2}$ $\frac{2}{3} * 3 = \frac{6}{3} = 2$ $\frac{2}{3} * \frac{3}{4} = \frac{6}{12} = \frac{1}{2}$ $9 + 1 = 10$ $6 + 1 + 2$ ⎯⎯⎯↑ $\frac{1}{2} + \frac{1}{2}$ ⎯⎯⎯⎯↑	Rename the mixed numbers as fractions. $3\frac{3}{4} = \frac{15}{4}$ $2\frac{2}{3} = \frac{8}{3}$ Multiply. $\frac{15}{4} * \frac{8}{3} = \frac{15 * 8}{4 * 3} = \frac{120}{12}$ Rename the product. $\frac{120}{12} = 10$

Multiply. Use one of the methods above or one of your own. Show your work.

1. $2\frac{3}{8} * \frac{1}{5} = $ _____

2. $3\frac{2}{5} * 2\frac{2}{3} = $ _____

3. $5\frac{3}{4} * 2\frac{5}{9} = $ _____

4. _____ $= 4\frac{6}{7} * \frac{3}{7}$

5. $4 * 6\frac{7}{10} = $ _____

6. $3\frac{11}{12} * 2\frac{1}{2} = $ _____

7. _____ $= 5\frac{7}{8} * 2\frac{1}{8}$

8. $4\frac{1}{6} * 3\frac{9}{4} = $ _____

Date _____ Time _____

Use the formulas below to help you find the area of each figure below.

Area of a rectangle	Area of a triangle	Area of a parallelogram
$A = b * h$	$A = \frac{1}{2} * b * h$	$A = b * h$

9.

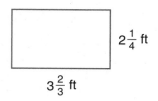

$2\frac{1}{4}$ ft

$3\frac{2}{3}$ ft

Area = _____ (unit)

10.

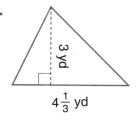

3 yd

$4\frac{1}{3}$ yd

Area = _____ (unit)

11.

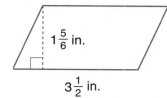

$1\frac{5}{6}$ in.

$3\frac{1}{2}$ in.

Area = _____ (unit)

12. Joan made a cubic box out of cardboard.
What is the area of all the cardboard she used?

Area = _____ (unit)

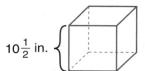

$10\frac{1}{2}$ in.

Try This

13. Lydia is putting photographs in an album. She does not like to leave more than $\frac{1}{3}$ of a page uncovered.

a. At the right are the dimensions of an album page and 4 photographs. If Lydia puts these 4 photographs on 1 page, what area of the page will be left uncovered?

_____

b. Is this amount more or less than $\frac{1}{3}$ of the total area of the page?

_____

	Height	Width
Album page	$6\frac{3}{4}''$	$9''$
Photograph 1	$4\frac{1}{8}''$	$2\frac{1}{2}''$
Photograph 2	$3\frac{5}{8}''$	$4''$
Photograph 3	$2\frac{1}{8}''$	$2\frac{3}{4}''$
Photograph 4	$2\frac{1}{8}''$	$2\frac{1}{8}''$

LESSON 4·7   **Math Boxes**

---

1. Rename each mixed number in simplest form.

   a. _____ $= 1\frac{17}{5}$

   b. $3\frac{46}{9} =$ _____

   c. $2\frac{35}{7} =$ _____

   d. _____ $= 1\frac{68}{8}$

   72

2. Subtract. Write the answer as a mixed number in simplest form.

   a. $8\frac{1}{4} - 2\frac{2}{3} =$ _____

   b. $4\frac{5}{6} - 2\frac{2}{3} =$ _____

   c. $6\frac{5}{8} - 2\frac{7}{10} =$ _____

   d. $5\frac{3}{4} - 3\frac{1}{8} =$ _____

   85   86

3. Multiply and divide mentally.

   a. $0.04 * 100 =$ _____

   b. $4{,}537 \div 10 =$ _____

   c. _____ $= 0.09 \div 100$

   d. _____ $= 1.0508 * 10{,}000$

   35 36
   40 41

4. Write the products in standard notation.

   a. _____ $= 49.1 * \frac{1}{10}$

   b. $12.5 * \frac{1}{100} =$ _____

   c. $3{,}825 * 10^{-2} =$ _____

   d. _____ $= 97.6 * 10^{-1}$

   8

5. Fill in the circle next to the best estimate for the quotient.

   $15\overline{)544.5}$

   Ⓐ 500

   Ⓑ 400

   Ⓒ 50

   Ⓓ 40

   42–45

6. Draw and label the following angle.

   $\angle MAL = 64°$

   SRB
   230 232

LESSON
4·8
Math Boxes

1. Compare. Write < or >.

   a. $\frac{7}{3}$ _____ $\frac{3}{7}$

   b. $\frac{5}{6}$ _____ $\frac{5}{7}$

   c. $\frac{2}{3}$ _____ $\frac{4}{9}$

   d. $\frac{5}{8}$ _____ $\frac{6}{13}$

SRB
75

2. Multiply. Write each answer in simplest form.

   a. $\frac{3}{8} * 6 =$ _____

   b. $2\frac{1}{7} * \frac{4}{5} =$ _____

   c. _____ $= \frac{8}{9} * 2\frac{1}{3}$

   d. $6\frac{2}{3} * 3\frac{3}{4} =$ _____

SRB
88 89

3. This spreadsheet shows students' times, in seconds, for 2 different runs.

   a. Calculate the mean running time for each student. Write the mean times in the appropriate cells.

   b. Using cell names, write a formula for calculating Keo's mean time.

   _____

	A	B	C	D
1	Student	Run 1	Run 2	Mean
2	Keo	23	21	
3	Carmine	20	19	
4	Ann	21.5	21	

SRB
142–144

4. Evaluate the following algebraic expressions for $k = \frac{3}{5}$.

   a. $k + 2\frac{3}{8}$ _____

   b. $7 - k$ _____

   c. $k - \frac{1}{2}$ _____

   d. $15 * k$ _____

SRB
83 84
89 242

5. Measure the angle to the nearest degree.

   m∠ TAP is _____ °

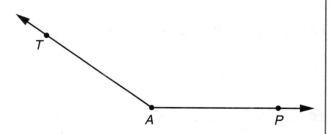

SRB
230 231

LESSON 4·8   **Fractions, Decimals, and Percents**

Math Message

During the baseball season, Sari got a hit
2 out of every 5 times she was at bat.

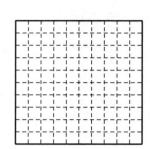

1. Shade $\frac{2}{5}$ of the square at the right.

2. How many hundredths are shaded? _____

3. $\frac{2}{5} = \frac{x}{100}$ $x =$ _____

You can rename some fractions as decimals by first renaming them as equivalent
fractions with 10 or 100 in the denominator.

Example 1:	**Example 2:**
$\frac{3}{5} = \frac{b}{10}$ $b = 6$	$\frac{3}{20} = \frac{d}{100}$ $d = 15$
If $\frac{6}{10} = 0.6$, then $\frac{3}{5} = 0.6$.	If $\frac{15}{100} = 0.15$, then $\frac{3}{20} = 0.15$.

Find the value of the variable. Use it to rename the fraction as a decimal.

4. $\frac{1}{4} = \frac{f}{100}$; $f =$ _____

$\frac{1}{4} = 0.$_____

5. $\frac{4}{5} = \frac{r}{100}$; $r =$ _____

$\frac{4}{5} = 0.$_____

6. $\frac{7}{20} = \frac{z}{100}$; $z =$ _____

$\frac{7}{20} = 0.$_____

7. $\frac{9}{2} = \frac{n}{100}$; $n =$ _____

$\frac{9}{2} = 4.$_____

Rename each fraction as a decimal.

8. $\frac{3}{4} = 0.$_____

9. $\frac{12}{25} = 0.$_____

10. $\frac{19}{50} = 0.$_____

11. $\frac{3}{2} = 1.$_____

Rename each decimal as a fraction in simplest form.

> **Examples:**
>
> $0.6 = \frac{6}{10} = \frac{3}{5}$ $0.32 = \frac{32}{100} = \frac{8}{25}$

12. $0.5 =$ _____

13. $0.25 =$ _____

14. $0.4 =$ _____

15. $0.65 =$ _____

16. $0.75 =$ _____

17. $0.46 =$ _____

18. $0.89 =$ _____

19. $0.36 =$ _____

Rename each fraction as a percent.

> **Examples:**
>
> $\frac{2}{5} = \frac{40}{100} = 40\%$ $\frac{9}{20} = \frac{45}{100} = 45\%$

20. $\frac{1}{4} = \frac{\boxed{}}{100} =$ _____%

21. $\frac{3}{5} = \frac{\boxed{}}{100} =$ _____%

22. $\frac{7}{10} = \frac{\boxed{}}{100} =$ _____%

23. $\frac{28}{50} = \frac{\boxed{}}{100} =$ _____%

24. $\frac{11}{20} = \frac{\boxed{}}{100} =$ _____%

25. $\frac{17}{25} = \frac{\boxed{}}{100} =$ _____%

26. $\frac{1}{20} = \frac{\boxed{}}{100} =$ _____%

27. $\frac{13}{10} = \frac{\boxed{}}{100} =$ _____%

Rename each percent as a fraction in simplest form.

> **Examples:**
>
> $40\% = \frac{40}{100} = \frac{2}{5}$ $34\% = \frac{34}{100} = \frac{17}{50}$

28. $25\% = \frac{\boxed{}}{100} = \frac{\boxed{}}{4}$

29. $20\% = \frac{\boxed{}}{100} = \frac{\boxed{}}{5}$

30. $30\% = \frac{\boxed{}}{100} = \frac{\boxed{}}{10}$

31. $80\% = \frac{\boxed{}}{100} = \frac{\boxed{}}{5}$

32. $75\% = \frac{\boxed{}}{100} = \frac{\boxed{}}{4}$

33. $95\% = \frac{\boxed{}}{100} = \frac{\boxed{}}{20}$

34. $120\% = \frac{\boxed{}}{100} = \frac{\boxed{}}{5}$

35. $150\% = \frac{\boxed{}}{100} = \frac{\boxed{}}{2}$

LESSON 4·9 Converting Between Decimals and Percents

Math Message

Fill in the blanks.

1. $0.36 = \dfrac{0.36 * 100}{100} = \dfrac{\boxed{}}{100} = $ _____ %

2. $0.7 = \dfrac{0.7 * 100}{100} = \dfrac{\boxed{}}{100} = $ _____ %

3. $0.09 = \dfrac{0.09 * 100}{100} = \dfrac{\boxed{}}{100} = $ _____ %

4. $4.602 = \dfrac{4.602 * 100}{100} = \dfrac{\boxed{}}{100} = $ _____ %

Rename each decimal as a percent.

5. 0.42 = _____

6. 0.08 = _____

7. 1.5 = _____

8. 7.36 = _____

Rename each percent as a decimal.

9. 23% = _____

10. 4% = _____

11. 314% = _____

12. 1,260% = _____

13. Divide using a calculator to fill in the table. Study the first problem.

Fraction	Decimal (Quotient)	Percent	Percent (to the nearest whole)
$\frac{1}{9}$	0.1111111111	11.11111111%	11%
$\frac{2}{3}$			
$\frac{3}{7}$			
$\frac{5}{9}$			
$\frac{7}{1}$			
$\frac{7}{8}$			

LESSON 4·9 Math Boxes

1. Write the following fractions in order from smallest to largest.

$\frac{7}{8}$ $\frac{1}{5}$ $\frac{99}{100}$ $\frac{6}{10}$ $\frac{5}{12}$

_____

SRB 75

2. Use the partial-quotients algorithm to divide the numerator by the denominator. Round the result to the nearest hundredth and rename the result as a percent.

$\frac{7}{12}$ = 0._____ = _____%

SRB 55–57

3. Fill in the missing numbers.

Fraction	Decimal	Percent
$\frac{3}{4}$		
	0.25	
		40%
		65%
	0.9	

SRB 59 60

4. Complete.

a. $\frac{1}{8}$ of 48 = _____

b. $\frac{5}{9}$ of 90 = _____

c. $\frac{2}{7}$ of 35 = _____

d. $\frac{3}{4}$ of 64 = _____

e. $\frac{4}{5}$ of 800 = _____

SRB 87

5. Use the formula $d = r * t$ (distance = rate $*$ time) to find the value of d when $r = 6\frac{1}{2}$ miles per hour and $t = 2\frac{1}{5}$ hours.

$d =$ _____ miles

SRB 246

6. Find the measure of $\angle N$ without measuring the angle.

m$\angle N$ is _____ $^\circ$

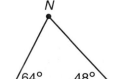

SRB 233

LESSON 4·10 **How to Draw a Circle Graph**

To draw a circle graph:

Step 1: Express each part of the data as a percent of the total.

Step 2: Use the Percent Circle on your Geometry Template to divide a circle and its interior (the total) into sectors whose sizes correspond to the percent form of the data.

1. Art, Barb, and Cyrus ran for sixth-grade class president. The election results are shown in the table below. Fill in the percent column. Then use your Percent Circle to draw a circle graph of the data.

Election Results		
Name	Number of Votes	Percent of Total Vote
Art	8	
Barb	10	
Cyrus	22	
Total	**40**	**100%**

Election Results

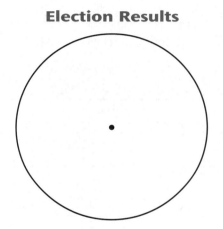

2. Doug, Bree, and Fred drove from Denver, Colorado, to Kansas City, Missouri. The number of hours each one drove is shown in the table below. Fill in the percent column. Then use your Percent Circle to draw a circle graph of the data.

Driving Times		
Name	Number of Hours	Percent of Total Time
Doug	5	
Bree	4.5	
Fred	3	
Total	**12.5**	**100%**

Driving Times

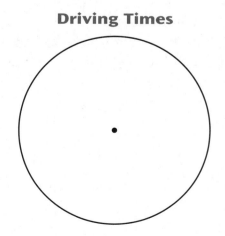

Date _____ Time _____

1. Compare. Write < or >.

a. $\frac{4}{5}$ _____ $\frac{5}{4}$

b. $\frac{7}{8}$ _____ $\frac{8}{9}$

c. $\frac{17}{18}$ _____ $\frac{5}{6}$

d. $\frac{1}{5}$ _____ $\frac{1}{9}$

SRB
75

2. Multiply. Write each answer in simplest form.

a. $\frac{4}{5} * \frac{7}{8} =$ _____

b. $\frac{11}{12} * \frac{5}{10} =$ _____

c. _____ $= 1\frac{3}{4} * 2\frac{1}{5}$

d. _____ $= 2\frac{1}{3} * 3\frac{1}{2}$

SRB
88 89

3. This spreadsheet shows Blaire's and Denise's scores on their first 3 science tests.

a. What is shown in cell C2?

_____

b. Calculate the values for cells E2 and E3 and enter them in the spreadsheet.

	A	B	C	D	E
1	Student	Test 1	Test 2	Test 3	Mean
2	Blaire	95	90	85	
3	Denise	80	75	100	

c. Fill in the circle next to the correct formula for Blaire's mean score for Tests 1, 2, and 3.

Ⓐ $\frac{(A2 + B2 + C3)}{3}$ Ⓑ $\frac{(B1 + B2 + B3)}{3}$ Ⓒ $\frac{(B2 + C2 + D2)}{3}$

SRB
142–144

4. Evaluate the following algebraic expressions for $w = 2\frac{1}{4}$.

a. $w + 1\frac{5}{8}$ _____

b. $9 - w$ _____

c. $w * 1\frac{1}{4}$ _____

d. $w - 1\frac{3}{8}$ _____

SRB
83 84
89 242

5. Measure the angle to the nearest degree.

m∠*DEN* is _____ °

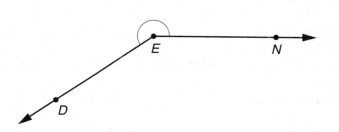

SRB
230 231

153

How Much Does Your Garbage Weigh?

LESSON 4·10

Composition of Garbage Generated, per Person per Day

Material[1]	1960	1980	2000	2005[2]
	Weight (lb)			
Paper and paperboard	0.91	1.32	1.77	1.64
Glass	0.20	0.36	0.24	0.24
Metals	0.32	0.35	0.36	0.35
Plastics	0.01	0.19	0.47	0.49
Other (rubber, textiles, wood)	0.24	0.44	0.68	0.71
Food wastes	0.37	0.32	0.46	0.52
Yard trimmings	0.61	0.66	0.46	0.57
Total garbage generated	**2.66**	**3.64**	**4.44**	**4.52**

Source: United States Environmental Protection Agency

[1]Included in table: household garbage and appliances; garbage from offices, businesses, restaurants, schools, hospitals, and libraries. Not included in table: car bodies, sludge, industrial and agricultural wastes.
[2]Estimated

1. The percent column for 1960 has been completed in the table below. Complete the percent column for 2000. Round each answer to the nearest percent.

Material	1960		2000	
	Weight (lb)	Percent of Total Weight	Weight (lb)	Percent of Total Weight
Paper and paperboard	0.91	34%	1.77	
Glass	0.20	8%	0.24	
Metals	0.32	12%	0.36	
Plastics	0.01	0% (< 0.004%)	0.47	
Other (rubber, textiles, wood)	0.24	9%	0.68	
Food wastes	0.37	14%	0.46	
Yard trimmings	0.61	23%	0.46	
Total garbage generated	**2.66**	**100%**	**4.44**	

How Much Does Your Garbage Weigh? *cont.*

2. Study the circle graph for the 1960 data. Then draw a circle graph for the 2000 data. Remember to graph the smallest sector first.

Garbage by Weight—1960

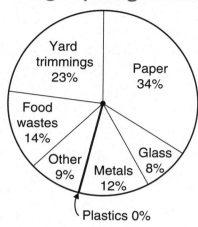

Yard trimmings 23%
Paper 34%
Food wastes 14%
Other 9%
Metals 12%
Glass 8%
Plastics 0%

Garbage by Weight—2000

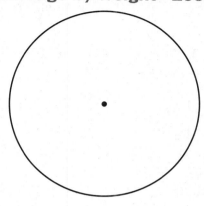

3. What material makes up the largest part
   of Americans' garbage, by weight?   _____

4. According to information for the year 2005, about how many pounds of garbage
   did a person generate, on average, in

   1 day? _____ 1 week? _____ 1 month (30 days)? _____

5. About how many months did it take a person
   to generate 1 ton of garbage in 2005?   _____

6. Describe the changes that took place in the composition of garbage
   from 1960 to 2000.

   _____

   _____

   _____

   _____

 **LESSON 4·10** **Multiplying Fractions and Mixed Numbers**

Solve.

1. $\frac{1}{4}$ of 16 _____

2. $\frac{6}{8}$ of 120 _____

3. 50% of 60 _____

4. $\frac{1}{4}$ of 2 _____

5. 25% of $\frac{2}{5}$ _____

6. $\frac{4}{6}$ of $\frac{4}{5}$ _____

Multiply. Write each answer in simplest form. If possible, write answers as mixed numbers or whole numbers.

7. $\frac{2}{5} * \frac{3}{8} =$ _____

8. $\frac{1}{3} * \frac{1}{2} =$ _____

9. $\frac{2}{3} * \frac{4}{5} =$ _____

10. $\frac{4}{9} * \frac{3}{8} =$ _____

11. $1\frac{2}{3} * 2\frac{1}{5} =$ _____

12. $3\frac{1}{4} * 2\frac{3}{4} =$ _____

13. $1\frac{2}{5} * 3 =$ _____

14. $2\frac{5}{6} * 2\frac{1}{2} =$ _____

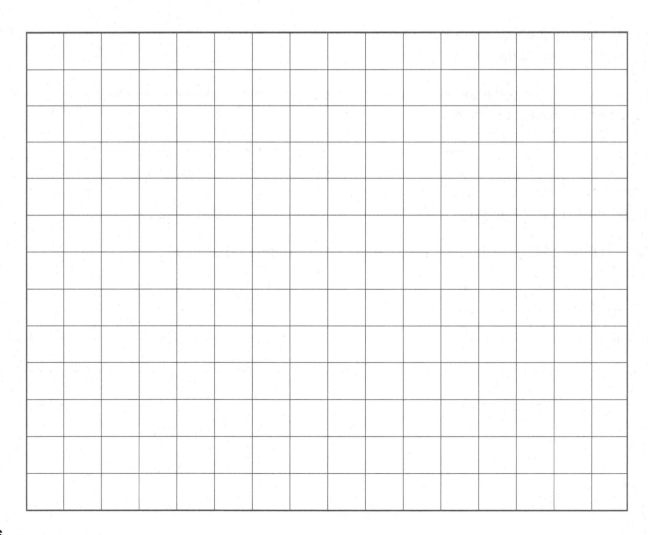

LESSON 4·11   Math Boxes

---

1. Write the following fractions in order from smallest to largest.

$\frac{1}{20}$ $\frac{6}{7}$ $\frac{9}{10}$ $\frac{1}{3}$ $\frac{5}{8}$

_____

SRB 75

---

2. Which of the following percents is equivalent to $\frac{5}{6}$? Choose the best answer.

⬭ 5.6%

⬭ 57%

⬭ 8.3%

⬭ 83%

SRB 55–57

---

3. Fill in the missing numbers.

Fraction	Decimal	Percent
$\frac{1}{5}$		
	0.625	
		60%
	0.9	
		5%

SRB 59 60

---

4. Complete.

a. $\frac{2}{5}$ of 75 = _____

b. $\frac{3}{8}$ of 24 = _____

c. $\frac{5}{6}$ of 48 = _____

d. $\frac{1}{2}$ of $2\frac{1}{2}$ = _____

e. $\frac{3}{12}$ of 32 = _____

SRB 87

---

5. Use the formula $i = \frac{c}{2.54}$ to convert centimeters to inches.

Evaluate the formula when $c = 76.2$.

$i =$ _____

76.2 cm = _____ in.

SRB 246

---

6. Find the measure of $\angle R$ without measuring the angle.

m$\angle R$ is _____°

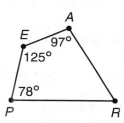

SRB 233

Date _____ Time _____

LESSON 4·11 Percent Problems

Math Message

A DVD that regularly costs $25 is on sale for 20% off.

1. What is the discount in dollars? _____

2. What is the sale price? _____

Solve Problems 3–5 using paper and pencil.

3. An electronic handheld game player that costs $120 is on sale for 25% off.

  a. What is the discount in dollars? _____

  b. What is the sale price? _____

4. A bank offers a simple interest rate of 6% per year.

  a. What is the interest on $100 at the end of 1 year? _____

  b. What is the interest on $500 at the end of 1 year? _____

  c. What is the interest on $1,000 at the end of 1 year? _____

5. The $2.50 price of a hot dog at a neighborhood baseball park is shared as follows:

  20% pays for the hot dog, bun, and fixings.

  20% pays the concession stand workers.

  10% is profit for the concession stand owner.

  10% pays for the electricity/fuel costs to cook and heat the hot dogs.

  40% pays for rental fees on the stand and cooking equipment.

  a. How much do the hot dog, bun, and fixings cost? _____

  b. How much goes to rental fees? _____

  c. How much does the concession stand owner get? _____

You may use a calculator to help you solve Problems 6–9.

6. A CD that regularly costs $14.95 is on sale for 20% off.

  a. What is the discount in dollars? _____

  b. What is the sale price? _____

158

LESSON 4·11 Percent Problems *continued*

7. The table below shows the results of a science literacy test given to 1,574 adults in 2001. Round to the nearest whole number to find the number of people who correctly answered each question.

Question	Percent Correct	Number of People
How long does it take Earth to go around the Sun: One day, one month, or one year? *Answer:* One year	54%	
True or false: Lasers work by focusing sound waves. *Answer:* false	45%	
Which travels faster, light or sound? *Answer:* light	76%	
True or false: The earliest humans lived at the same time as the dinosaurs. *Answer:* false	48%	

Source: Science and Engineering Indicators—2002

8. In 1998, there were about 14.5 million high school students in the United States. The number of high school students is expected to increase about 7% between 1998 and 2010.

a. About how many more high school students will there be in 2010?

_____

b. About how many high school students will there be in all in 2010?

_____

Source: National Center for Education Statistics (NCES)

9. Suppose elementary school students make up about 15% of the U.S. population. About how many elementary school students would you expect to live in a town that has a population of 50,000?

_____

LESSON 4·11   Divide to Rename Fractions

SRB 55–57

For each fraction, use the partial-quotients division algorithm to divide the numerator by the denominator. Round the result to the nearest hundredth and rename it as a percent.

1. $\frac{4}{7}$ = 0._____ = _____%   **2.** $\frac{7}{12}$ = 0._____ = _____%   **3.** $\frac{9}{16}$ = 0._____ = _____%

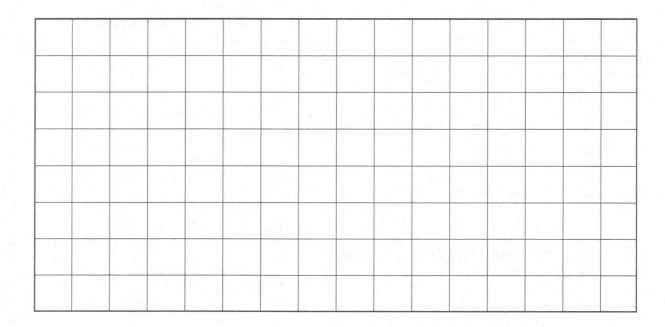

4. $\frac{5}{16}$ = 0._____ = _____%   **5.** $\frac{1}{15}$ = 0._____ = _____%   **6.** $\frac{11}{12}$ = 0._____ = _____%

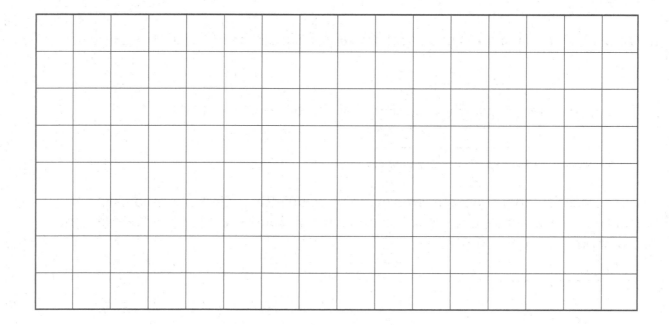

LESSON 4·12   Math Boxes

1. The table below shows the results of a Bureau of the Census study of how people get to work. Use the Percent Circle to draw a circle graph of the results.

Method of Transportation	Percent of People
Drive alone	75.1%
Carpool	13.2%
Walk or work at home	5.3%
Take public transportation or other	6.4%

How People Get to Work

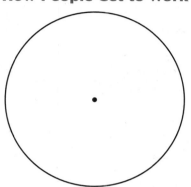

SRB
146

2. Draw and label the following angles.

   a. ∠LAG = 72°

   b. ∠AND = 125°

SRB
230 232

3. Measure each angle to the nearest degree.

   a.
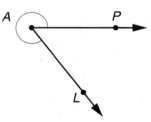

   b.
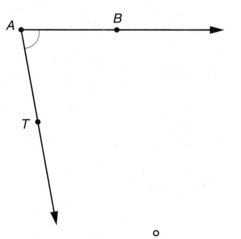

   m∠LAP is _____ °

   m∠TAB is _____ °

SRB
230 231

161

LESSON 5·1 # Measuring and Drawing Angles

Math Message

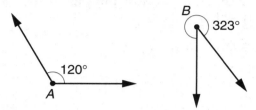 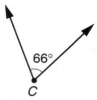 

Which of the above angles is

1. a right angle? ∠ _____

2. an acute angle? ∠ _____

3. a reflex angle? ∠ _____

4. an obtuse angle? ∠ _____

5. a straight angle? ∠ _____

---

6. Use your full-circle protractor to measure each angle to the nearest degree.

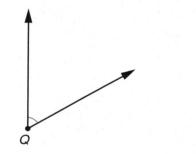 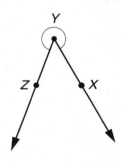

    a. ∠Q measures about _____.   **b.** ∠XYZ measures about _____.

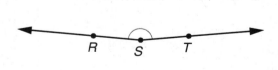 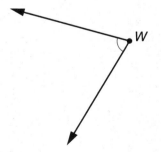

    c. ∠RST measures about _____.   **d.** ∠W measures about _____.

LESSON 5·1 **Measuring and Drawing Angles** *continued*

7. Use your half-circle protractor to measure each angle to the nearest degree.

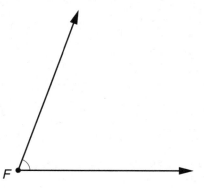

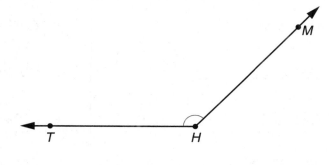

a. ∠F measures about _____.

b. ∠THM measures about _____.

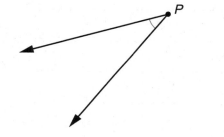

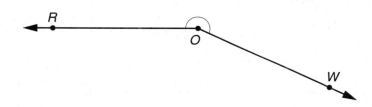

c. ∠P measures about _____.

d. ∠ROW measures about _____.

8. Draw and label the following angles. Use your half-circle protractor.

a. ∠DOR: 43°

b. ∠CAN: 165°

LESSON 5·1 *Percent of* Problems

Solve mentally.

1. 10% of 20 = _____

2. 20% of 50 = _____

3. 25% of 320 = _____

4. 40% of 25 = _____

5. 50% of 36 = _____

6. 15% of 200 = _____

7. 75% of 120 = _____

8. 60% of 45 = _____

9. Choose one problem from Problems 1–8. Explain the strategy you used to solve that problem.

_____

_____

_____

_____

Solve. You may use your calculator.

10. $33\frac{1}{3}$% of 48 = _____

11. 45% of 72 = _____

12. 68% of 19 = _____

13. 8% of 30 = _____

14. Choose one problem from Problems 10–13. Explain a strategy you could use to mentally solve that problem.

_____

_____

_____

_____

Date _____   Time _____

1. Estimate the degree measure of ∠*CAT*.

Estimate _____°

Then use your half-circle protractor to measure ∠*CAT* to the nearest degree.

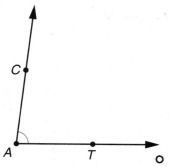

∠*CAT* measures about _____ .

SRB 230–232

2. Use your half-circle protractor to draw an angle measuring 145°. Label it ∠*HJK*.

SRB 232

3. Solve mentally.

a. $\frac{5}{7}$ of 28 = _____

b. 2% of 200 = _____

c. $\frac{3}{5}$ of 45 = _____

d. $8\frac{1}{2}$% of 100 = _____

SRB 49 50 87

4. Rewrite each fraction pair using a common denominator.

a. $\frac{3}{8}$ and $\frac{2}{3}$ _____ and _____

b. $\frac{6}{9}$ and $\frac{4}{12}$ _____ and _____

c. $\frac{2}{10}$ and $\frac{15}{25}$ _____ and _____

d. $\frac{5}{1}$ and $\frac{6}{3}$ _____ and _____

SRB 79

5. Estimate each product. Then multiply and simplify.

a. $4\frac{5}{6} * 18\frac{3}{4}$ Estimate _____

Actual _____

b. $12\frac{5}{8} * 3\frac{1}{3}$ Estimate _____

Actual _____

SRB 90

6. Insert parentheses to make each sentence true.

a. 42 / 12 / 2 = 7

b. 18 + 5 * 2 = 46

c. 125 − 25 * 5 = 0

d. 72 / 8 + 4 / 6 = $9\frac{2}{3}$

SRB 247

**LESSON
5·2** **Angle Relationships**

Math Message

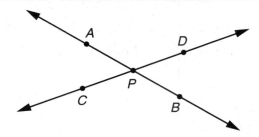

Measure angles *APC, DPB, APD,* and *CPB.* Write the measures below.

(*Note:* m∠*APC* is short for *the measure of angle* APC.)

1. m∠*APC* = _____   m∠*DPB* = _____

2. m∠*APD* = _____   m∠*CPB* = _____

3. What do you notice about the angle measures?

   _____

   _____

   _____

Angles are sometimes named with lowercase letters. Find the measures
of the angles indicated in Problems 4–13. Do *not* use a protractor.

4.
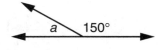

m∠*a* = _____

5.
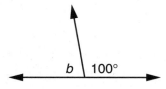

m∠*b* = _____

6.
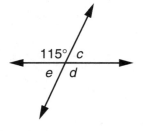

m∠*c* = _____

m∠*d* = _____

m∠*e* = _____

7.
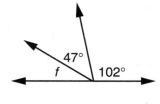

m∠*f* = _____

166

**LESSON
5·2**   **Angle Relationships** *continued*

8.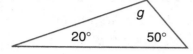

m∠g = _____

9.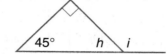

m∠h = _____

m∠i = _____

Reminder: The symbol ⌐ means that the angle is a right angle.

10.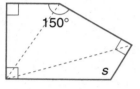

m∠s = _____

11.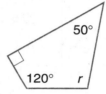

m∠r = _____

12.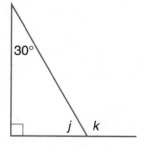

m∠j = _____

m∠k = _____

13.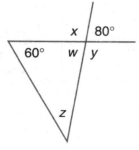

m∠w = _____

m∠x = _____

m∠y = _____

m∠z = _____

LESSON 5·2   Math Boxes

1. Use a protractor to find the angle measures in the regular hexagon *ABCDEF*. Then find the sum of the measures.

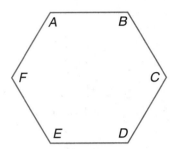

Sum of the
angle measures = _____ °

SRB 231

2. Name 1 pair of vertical angles and 1 pair of adjacent angles.

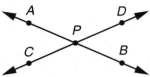

vertical angles: _____ and _____

adjacent angles: _____ and _____

SRB 163

3. A leather coat that regularly costs $325 is on sale for 20% off. What is the sale price?

Sale Price _____

SRB 49 50

4. Write in standard notation.

a. $(3 * 10^2) + (5 * 10^0) + (9 * 10^{-3})$

_____

b. 0.7 million _____

c. $(4 * 10^0) + (2 * 10^{-2}) + (7 * 10^{-4})$

_____

d. 1.5 thousand _____

5. Add or subtract.

a. $15 + (-18) =$ _____

b. $14 - 16 =$ _____

c. $28 + (-30) =$ _____

d. $0 - 45 =$ _____

SRB 95 96

6. Compare using $<$, $>$, or $=$.

a. $8 + (3 + 10)$ _____ $(8 + 3) + 10$

b. $12 \div 3$ _____ $3 \div 12$

c. $15 * 25$ _____ $25 * 15$

d. $20 - (15 - 3)$ _____ $(20 - 15) - 3$

SRB 9 104

LESSON 5·3

Degree Measures of Sectors

Math Message

1. Find the measures of the following angles in circle O without using a protractor.

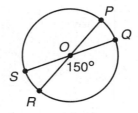

   a. m∠QOR = 150°

   b. m∠POS = _____°

   c. m∠POQ = _____°

   d. m∠SOR = _____°

2. Find the sum of the angle measures in Problem 1.

   150° + m∠POS + m∠POQ + m∠SOR = _____°

---

3. Connie, Josh, and Manuel were running for student council representative. The table below shows the number of votes that each candidate received. Complete the table.

Candidate	Number of Votes Received	Fraction of Votes Received	Percent of Votes Received
Connie	7	$\frac{7}{25}$	28%
Josh	6		
Manuel	12		
Total	25	$\frac{25}{25}$	100%

4. Use the percents from the table above to calculate the degree measure of the sector representing each candidate.

Candidate	Percent of Votes Received	Degree Measure of Sector (to nearest degree)
Connie	28%	0.28 * 360 = 100.8° ≈ 101°
Josh		
Manuel		
Total	100%	360°

LESSON 5·3 Drawing Circle Graphs with a Protractor

Mr. Li surveyed the students in his class to find out what kinds of pets they owned and how many of each kind they had. The results are shown in the first two columns of the table below.

Kind of Pet	Number of Pets	Fraction of Total Number of Pets	Decimal Equivalent (to nearest thousandth)	Percent of Total Number of Pets	Degree Measure of Sector
Dog	8	$\frac{8}{24}$	0.333	$33\frac{1}{3}\%$	$\frac{1}{3}$ * 360° = _120°_
Cat	6				_____ * 360° = _____
Guinea pig or hamster	3				_____ * 360° = _____
Bird	3				_____ * 360° = _____
Other	4				_____ * 360° = _____

1. Complete the table above.
   Study the first row.

2. At the right, or on a separate sheet of paper, use a compass and a protractor to make a circle graph of the data in the table. If you need to, tape your completed circle graph on this page. Write a title for the graph.

LESSON 5·3 **Drawing Circle Graphs with a Protractor** *cont.*

SRB
59 60
147

3. Sixth-grade students at Hawthorn School took a survey about after-school activities. Students answering the survey named the activity on which they spent the most time after school. The results are shown in the table below. Complete the table.

Activity	Number of Students	Fraction of Students	Decimal Equivalent	Percent of Students (to nearest percent)	Size of Sector
Music	12				
Math Club	28				
Art	5				
Sports	8				
Computers	3				
None	4				

4. In the space below, or on a separate sheet of paper, use a compass and a protractor to make a circle graph of the data in the table. If you need to, tape your completed circle graph on this page. Write a title for the graph.

LESSON 5·3 — Calculating Sale Price

Study each method for calculating sale price.

Two-Step Method	One-Step Method
Regular Price: $35.50 Discount: 20% Find the sale price.	Regular Price: $35.50 Discount: 20% Find the sale price.
Step 1: Find the discount in dollars: 20% of $35.50. $0.2 * \$35.50 = \$7.10$	The discount is 20%, so the amount of the regular price that remains is 100% − 20%, or 80% of the regular price.
Step 2: Subtract the discount amount from the regular price. Regular Price − Discount = Sale Price $\$35.50 - \$7.10 = \$28.40$	Find 80% of $35.50. 80% of $35.50 = $0.8 * \$35.50 = \$28.40$ The sale price is $28.40.
The sale price is $28.40.	

Use either method to find the sale price.

1. Regular Price: $99.00
Discount: 30%
Sale Price: _____

2. Regular Price: $45.00
Discount: 15%
Sale Price: _____

3. Regular Price: $435.00
Discount: 5%
Sale Price: _____

4. Regular Price: $348.50
Discount: 20%
Sale Price: _____

5. Regular Price: $4,380
Discount: 18%
Sale Price: _____

6. Regular Price: $25,125
Discount: 12%
Sale Price: _____

Date _____ Time _____

LESSON 5·3 Math Boxes

1. Estimate the degree measure of ∠QRS.

Estimate _____ °

Then use your full-circle protractor to measure ∠QRS to the nearest degree.

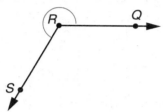

∠QRS measures about _____ °.

SRB
230–232

2. Use your half-circle protractor to draw an angle measuring 330°. Label it ∠NOP.

SRB
232

3. Solve mentally.

a. $\frac{2}{9}$ of 45 = _____

b. $33\frac{1}{3}$% of 18 = _____

c. $\frac{4}{5}$ of 100 = _____

d. 75% of 32 = _____

SRB
49 50
87

4. Rewrite each fraction pair using a common denominator.

a. $\frac{3}{10}$ and $\frac{8}{25}$ _____ and _____

b. $\frac{7}{15}$ and $\frac{19}{45}$ _____ and _____

c. $\frac{4}{5}$ and $\frac{3}{9}$ _____ and _____

d. $\frac{6}{27}$ and $\frac{7}{54}$ _____ and _____

SRB
79

5. Marta has $2\frac{3}{4}$ rolls of ribbon. If there are $3\frac{1}{3}$ yards of ribbon on each roll, about how many yards of ribbon does Marta have? Circle the best estimate.

A. $\frac{1}{2}$ yard

B. 6 yards

C. $6\frac{1}{2}$ yards

D. 9 yards

SRB
90

6. Insert parentheses to make each sentence true.

a. 20 − 16 / 2 * 15 = 30

b. 12 * 6 − 22 / 5 = 10

c. 72 / 8 + 4 / 6 = 1

d. 95 − 10 / 3 + 2 = 93

SRB
247

173

LESSON 5·4   The Coordinate Grid

1. Plot and label the following points on the coordinate grid below. The first one has been done for you.

   A: (3,1)   B: (2,5)   C: $(-4,-\frac{1}{2})$ D: (−6,−9)   E: (0,3)   F: (−2.5,0)

2. Write the ordered number pair for each of the following points shown on the coordinate grid below.

   G: (_____,_____)   H: (_____,_____)   I: (_____,_____)

   J: (_____,_____)   K: (_____,_____)   L: (_____,_____)

   M: (_____,_____)   N: (_____,_____)

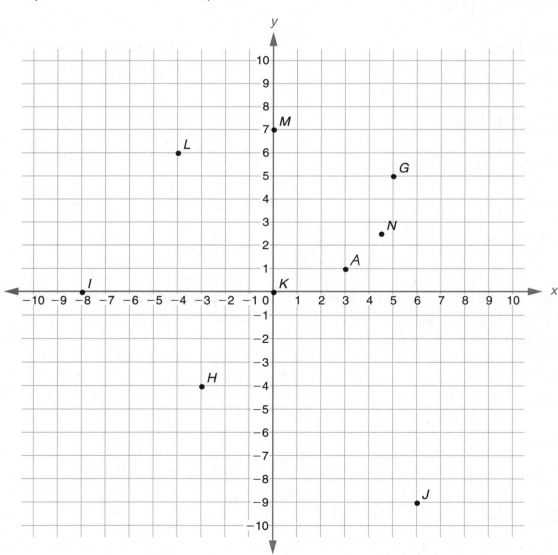

LESSON 5·4 Polygons on a Coordinate Grid

The names of polygons consist of letters that name the vertices, written in consecutive order. For example, the square at the right may be named square *ABCD*, *BCDA*, *CDAB*, or *DABC*.

The points shown on the grid below represent vertices of polygons. One or two vertices are missing for each polygon. Plot and name the missing vertices on the grid and then draw each polygon. List the number pairs for the missing vertices.

1. Scalene triangle *ABC*

   C: (_____, _____)

2. Right triangle *DEF*, which is also an isosceles triangle

   F: (_____, _____)

3. Square *GHIJ*

   I: (_____, _____)

   J: (_____, _____)

4. Rectangle *KLMN*, with $\overline{LM} = 2 * \overline{KL}$

   M: (_____, _____)

   N: (_____, _____)

5. Isosceles triangle *OPQ*, with $\overline{OP}$ the longest side

   Q: (_____, _____)

6. Parallelogram *RSTU*

   U: (_____, _____)

7. Rhombus *VWXY*

   W: (_____, _____)

   Y: (_____, _____)

8. Kite *A'D'P'R'*

   P': (_____, _____)   *R':* (_____, _____)

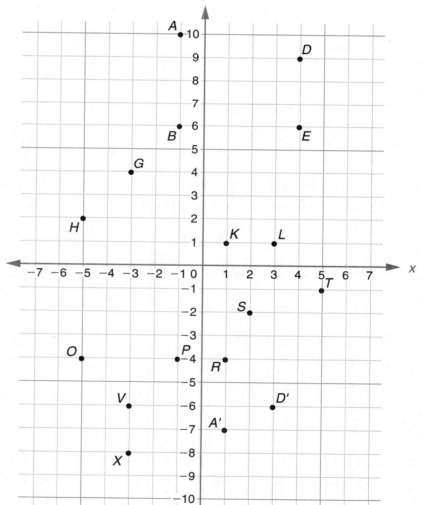

Date _____ Time _____

The **midpoint** of a line segment is the point halfway between the endpoints of the segment.

 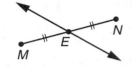 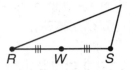

C is the midpoint of $\overline{AB}$.   *E* is the midpoint of $\overline{MN}$.   *W* is the midpoint of $\overline{RS}$.

1. Find the midpoint for each line segment shown and mark it on the grid.

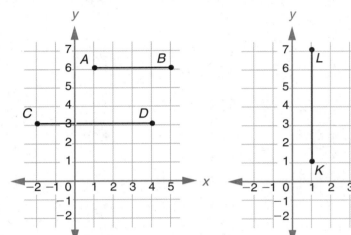 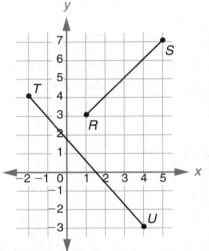

2. Find the endpoints and midpoint for each of the following line segments from Problem 1 above. Study the first one, which has been done for you.

	Endpoints		Midpoint
a. $\overline{AB}$	(_1_ , _6_)	(_5_ , _6_)	(_3_ , _6_)
b. $\overline{CD}$	(___ , ___)	(___ , ___)	(___ , ___)
c. $\overline{KL}$	(___ , ___)	(___ , ___)	(___ , ___)
d. $\overline{MN}$	(___ , ___)	(___ , ___)	(___ , ___)
e. $\overline{RS}$	(___ , ___)	(___ , ___)	(___ , ___)
f. $\overline{TU}$	(___ , ___)	(___ , ___)	(___ , ___)

3. Look for a pattern in your answers to Problem 2. If you know the coordinates of the endpoints of a line segment, how can you find the coordinates of the midpoint of the segment without plotting the line segment on a coordinate grid?

_____

_____

_____

LESSON 5·4 Math Boxes

1. Find the sum of the angle measures in the regular pentagon below without using a protractor.

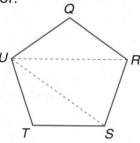

Sum of the angle measures = _____ °

2. Which angles below are adjacent angles? Circle the best answer.

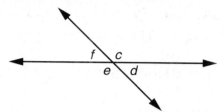

A $\angle f$ and $\angle c$ B $\angle f$ and $\angle d$

C $\angle c$ and $\angle e$ D $\angle d$ and $\angle f$

3. A television that regularly costs $799.00 is on sale for 15% off. What is the sale price of the television?

Sale Price _____

4. Write each number in standard notation.

a. $7.95 * 10^{-3}$ _____

b. 1.6 billion

_____

c. $(8 * 10^1) + (3 * 10^{-1}) + (7 * 10^{-3})$

_____

d. $196 * 10^{-2}$ _____

5. Add or subtract.

a. $96 + (-108) =$ _____

b. $37 - 85 =$ _____

c. $4 + (-4\frac{1}{2}) =$ _____

d. $-7 + 9 - 5 =$ _____

6. Find each missing value.

a. $1.8 + 2.5 = 2.5 + x$

$x =$ _____

b. $\frac{1}{3} * (5 + 4) = (5 + 4) * w$

$w =$ _____

LESSON 5·5 Isometry Transformations

Math Message

Translations (slides), reflections (flips), and **rotations (turns)** are basic transformations that can be used to move a figure from one place to another without changing its size or shape.

1. Study each transformation shown below.

Translation (Slide)	Reflection (Flip)	Rotation (Turn)
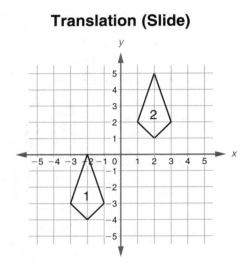	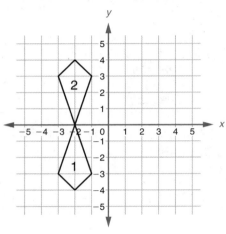	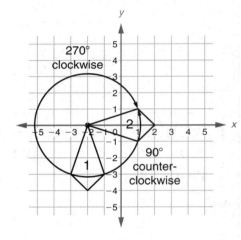

A **translation** (slide) moves each point of a figure a certain distance in the same direction.

A **reflection** (flip) of a figure gives its mirror image over a line.

A **rotation** (turn) moves a figure around a point.

2. Identify whether the preimage (1) and image (2) are related by a translation, a reflection, or a rotation. Record your answer on the line below each coordinate grid.

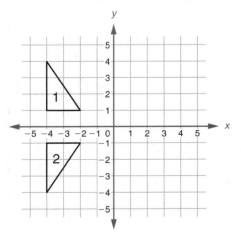	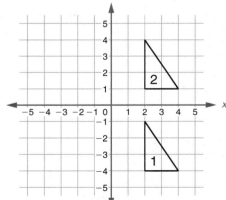	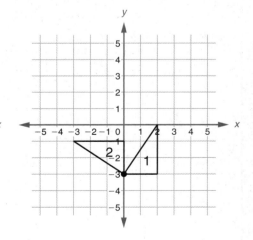

a. _____   b. _____   c. _____

LESSON 5·5

Translations

SRB
180–181

Example:

Translate quadrangle *ABCD* 6 units
to the right and 5 units up.

Plot and label the vertices of the
image that would result from
the translation.

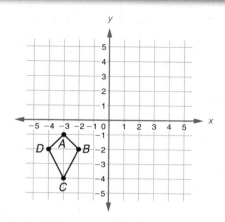

Plot and label the vertices of the image that would result from each translation.

1. Translate triangle *DEF* 4 units
to the right and 3 units down.

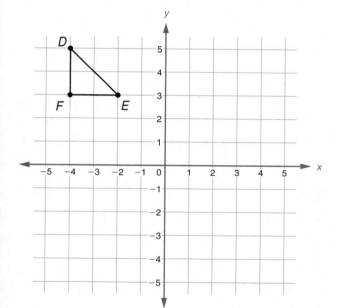

2. Translate pentagon *LMNOP*
0 units to the right and 7 units down.

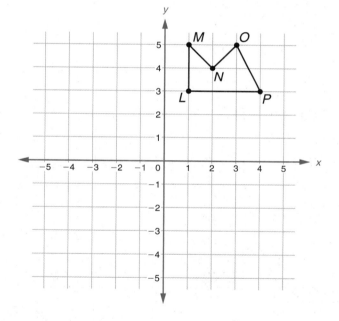

Try This

3. Square *WXYZ* has the following vertices:
$W(-3,-2)$, $X(-1,-2)$, $Y(-1,-4)$, $Z(-3,-4)$

Without graphing the preimage, list the vertices of image *W′X′Y′Z′* resulting from
translating each vertex 3 units to the right and 2 units up.

W' (_____,_____); X' (_____,_____); Y' (_____,_____); Z' (_____,_____)

LESSON 5·5   **Reflections**

Reflect each figure over the indicated axis or line of reflection. Then plot and label the vertices of the image that results from the reflection. Use a transparent mirror to check your placement of each image.

1. Reflect triangle *TAM* over the *x*-axis.

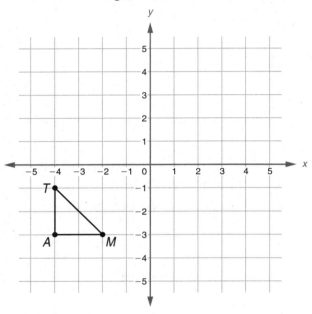

2. Reflect rectangle *STUV* over the *y*-axis.

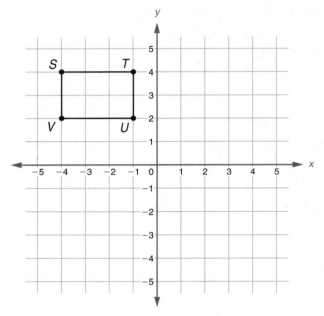

3. Reflect triangle *PQR* over the *y*-axis.

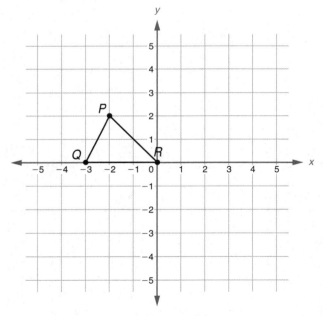

4. Reflect square *DEFG* over line *m*.

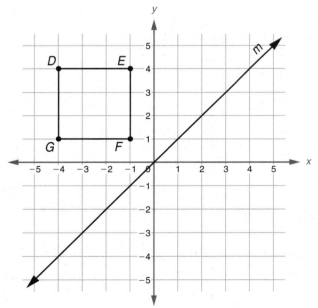

LESSON 5·5   **Rotations**

Rotate each figure around the point in the direction given. Then plot and label the vertices of the image that results from that rotation.

1. Rotate triangle *XYZ* 180° clockwise about Point *X*.

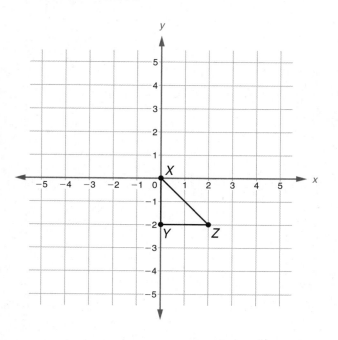

2. Rotate quadrangle *BCDE* 90° counterclockwise about the origin.

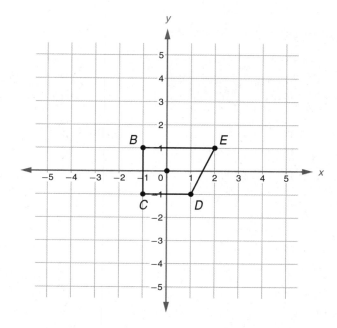

3. Rotate parallelogram *MNOP* 90° counterclockwise about point *M*.

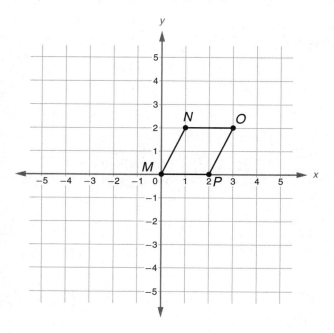

4. Rotate triangle *SEV* 270° clockwise about (0,2).

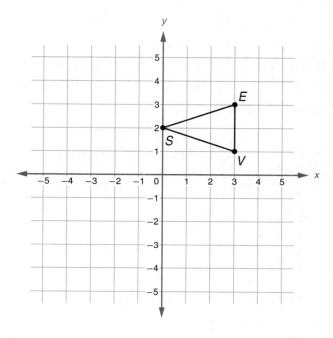

LESSON 5·5 Making a Circle Graph with a Protractor

1. One way to convert a percent to the degree measure of a sector is to multiply 360° by the decimal equivalent of the percent.

Example: What is the degree measure of a sector that is 55% of a circle?
$$55\% \text{ of } 360° = 0.55 * 360° = 198°$$

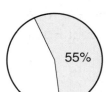

55%

Complete the table below.

Percent of Circle	Decimal Equivalent	Degree Measure of Sector
40%		$0.4 * 360° = $ _____
90%		
65%		
5%		
1%		

2. The table below shows the elective courses taken by a class of seventh graders. Complete the table. Then, in the space to the right, use a protractor to make a circle graph to display the information. Do not use the Percent Circle. (*Reminder:* Use a fraction or a decimal to find the degree measure of each sector.) Write a title for the graph.

Course	Number of Students	Fraction of Students	Decimal Equivalent	Percent of Students	Degree Measure of Sector
Music	6				
Art	9				
Computers	10				
Photography	5				

LESSON 5·5   **Math Boxes**

1. Rectangle *MNOP* has sides parallel to the axes. What are the coordinates of point *O*?

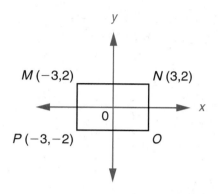

M (−3,2)   *N* (3,2)

P (−3,−2)   *O*

Coordinates of point *O*: (_____, _____)

SRB
234

2. Solve mentally.

   a. 20% of 50 = _____

   b. $\frac{3}{8}$ of 48 = _____

   c. _____ = 75% of 64

   d. _____ = $\frac{5}{9}$ of 90

SRB
49 50
87

3. a. Draw a line segment that is $3\frac{5}{16}$ inches long.

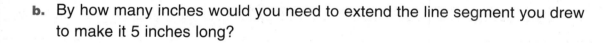

   b. By how many inches would you need to extend the line segment you drew to make it 5 inches long?

   _____

SRB
85 86

4. Multiply or divide.

   a. $45 * \frac{1}{5}$ = _____

   b. $60 \div 4$ = _____

   c. _____ = $56 * \frac{1}{8}$

   d. _____ = $108 \div 12$

SRB
88

5. Find the value that makes each number sentence true.

   a. $32 + n = 52$ $n =$ _____

   b. $y − 15 = 20$ $y =$ _____

   c. $(2 * m) + 5 = 17$ $m =$ _____

SRB
242 243

183

LESSON 5·6 Congruent Figures

Math Message

Carefully examine the figures in examples a–f.

The following pairs of figures are **congruent** to each other.

The following pairs of figures are *not* congruent to each other.

a.

d.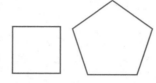

b.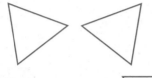

e.

c.

f.

1. Write a definition of *congruent polygons.* Then compare your definition to the definition on page 178 in the *Student Reference Book*.

_____

Line segments are **congruent** if they have the same length.	Angles are **congruent** if they have the same degree measure.

2. Draw wavy lines to connect each pair of congruent line segments below. Use a ruler to measure line segments if needed.

3. Draw wavy lines to connect each pair of congruent angles below. Use a protractor to measure angles if needed.

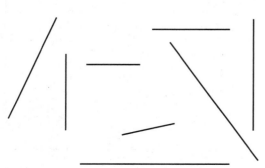

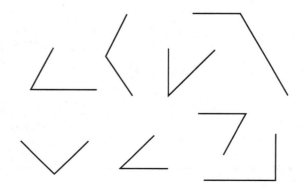

LESSON 5·6

Constructing Congruent Figures

You may use any of your construction tools—ruler, compass, protractor, or Geometry Template—to complete the constructions below. Do not trace.

1. Draw a triangle that is congruent to triangle *RST*.

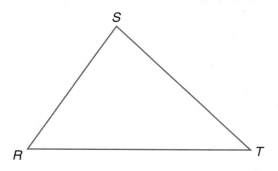

2. Draw a line segment *BC* so that $\overline{BC}$ is congruent to $\overline{AB}$ and the measure of angle *ABC* is 45°.

3. The plan below is for a paper cone with a glue tab. All dimensions and angle measures are shown. Draw a congruent copy of the plan on a separate sheet of paper. Cut out your copy and tape it on top of the plan below.

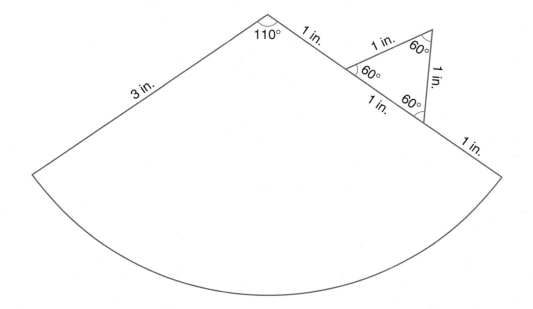

185

LESSON 5·6 **Rough Sketches and Accurate Drawings**

Roberta wants to build a bird feeder. She has found a plan for a feeder in a library book.

Roberta's rough sketch for one side of the feeder is shown below. The labels for the lengths and angle measures are correct, but the scale of the drawing is not accurate.

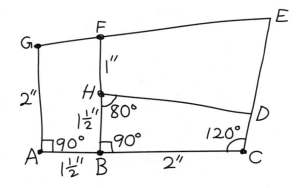

In the space below, make an accurate drawing of Roberta's rough sketch. Mark and label points A, B, C, D, E, F, G, and H on your drawing.

LESSON 5·6 Math Boxes

1. Rotate rhombus *BDFH* 90° counterclockwise about point *F* (0,0).
   Then plot and label the vertices of the image that results from that rotation.

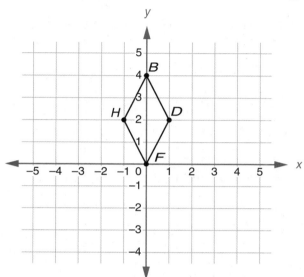

$B' = ($ _____ , _____ $)$ $D' = ($ _____ , _____ $)$

$F' = ($ _____ , _____ $)$ $H' = ($ _____ , _____ $)$

SRB 180 234

2. Use the partial-quotients algorithm to divide the numerator by the denominator. Round the result to the nearest hundredth and rename the result as a percent.

$\frac{9}{16} = 0.$ _____ $=$ _____ %

SRB 55–57

3. Add, subtract, or multiply.

   a. $2\frac{3}{6} + 3\frac{1}{2} =$ _____

   b. $4 - 2\frac{3}{5} =$ _____

   c. $2\frac{1}{4} - 1\frac{2}{3} =$ _____

   d. $\frac{2}{3} * \frac{4}{5} =$ _____

SRB 84–86 89

4. Write a number sentence for each word sentence. Then tell whether the number sentence is true or false.

Word Sentence	Number Sentence	True or False?
Five times eight is equal to 45.		
15 is greater than 2 less than 10.		
If 72 is divided by the square of 3, the result is 8.		

SRB 241–243

LESSON 5·7 Constructing Line Segments

Use only a compass, a straightedge, and a sharp pencil for the constructions below. Use rulers and protractors only to check your work. Do not trace.

1. Copy this line segment. Label the endpoints of your copy A' and B'. (These symbols are read A *prime* and B *prime*.)

A •————————————————————• B

2. Construct a line segment twice as long as $\overline{CD}$. Label the endpoints of your segment C' and D'.

C •————————————• D

3. Construct a line segment as long as $\overline{EF}$ and $\overline{GH}$ together. Label the endpoints of your segment E' and H'.

E •————————————————• F   G •————————————————————————————• H

Try This

4. Construct a segment with a length equal to the length of $\overline{IJ}$ minus the length of $\overline{KL}$. Label the endpoints of your segment I' and K'.

I •————————————————————————• J

K •————————————• L

LESSON 5·7 Constructing Triangles

Use only a compass, a straightedge, and a sharp pencil for the constructions below.
Use rulers and protractors only to check your work. Do not trace.

Make your constructions on another sheet of paper. If your compass has a sharp point,
work on top of a piece of cardboard or a stack of several sheets of paper. When you
are satisfied with a construction, cut it out and tape it onto this page.

1. Copy triangle *ABC*. Label the vertices of your copy *A′*, *B′*, and *C′*.

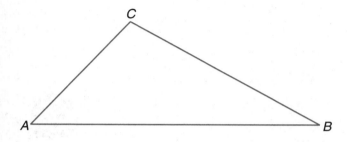

2. Construct a triangle with each side
   the same length as $\overline{DE}$.

3. Use a ruler to draw a line segment
   2 inches long and another line segment
   3 inches long. Then use a compass and a
   straightedge to construct a triangle with sides
   2 inches, 2 inches, and 3 inches long.

Try This

4. Is it possible to draw a triangle with sides 3 inches, 3 inches, and 7 inches long? _____

   Explain. _____

   _____

LESSON 5·7   Math Boxes

1. If point *A* (0,5) is reflected over the *x*-axis, what are the coordinates of *A'*?

   Fill in the circle next to the best answer.

   Ⓐ (5,0)

   Ⓑ (−5,0)

   Ⓒ (5,5)

   Ⓓ (0,−5)

SRB
180 234

2. Solve mentally.

   a. 40% of 55 = _____

   b. $\frac{5}{6}$ of 72 = _____

   c. _____ = 50% of $2\frac{1}{2}$

   d. _____ = $\frac{4}{5}$ of 100

SRB
49 50
88

3. **a.** Draw a line segment that is 8.6 cm long.

   b. By how many centimeters would you need to extend the line segment you drew to make it 10 centimeters long?

   _____

SRB
209

4. Multiply or divide.

   a. $63 * \frac{1}{9}$ = _____

   b. $81 \div 9$ = _____

   c. _____ = $140 * \frac{1}{7}$

   d. _____ = $180 \div 6$

SRB
88

5. Find the value that makes the number sentence true.

   a. $\frac{x}{12} = 12$ $x =$ _____

   b. $3 - (3 * b) = 0$ $b =$ _____

   c. $7 * (8 - n) = 21$ $n =$ _____

SRB
241–243

Date _____   Time _____

LESSON 5·8 Math Boxes

1. Without using a protractor, find the degree measure of each angle listed below.

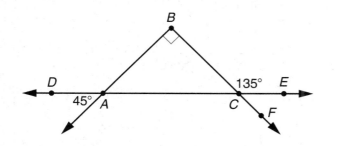

m∠*BAC* = _____

m∠*BAD* = _____

m∠*BCA* = _____

m∠*ECF* = _____

SRB 163

2. The instruments and number of players in the Playing Protractors rock band are shown in the table below. Complete the table. Then use a protractor to make a circle graph of the information. Title the graph.

Type of Instrument	Number of Players	Percent of Total	Degree Measure of Sector
Brass	7		
Keyboard	1	5%	18°
Percussion	3		
Strings	5		
Woodwind	4		
TOTAL	20	100%	360°

_____

_____

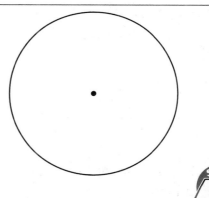

SRB 147

3. Find the missing value that makes the number sentence true.

a. $y + \frac{2}{3} = 5$ $y =$ _____

b. $t - 4\frac{1}{2} = 4\frac{1}{2}$ $t =$ _____

c. $3\frac{5}{8} + 2\frac{7}{12} = w$ $w =$ _____

SRB 84–86

4. Draw a line segment that is $1\frac{1}{2}$ inches long.

How many $\frac{1}{4}$-inch segments are in $1\frac{1}{2}$ inches?

_____

SRB 91

191

LESSON 5·8   Compass-and-Straightedge Constructions

Use only a compass, a straightedge, and a sharp pencil. Use rulers and protractors only to check your work. Do not trace. Make your constructions on another sheet of paper. When you are satisfied with a construction, cut it out and tape it onto this page.

1. Copy this angle. When you are finished, check your work with a protractor.

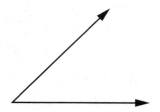

2. Copy this quadrangle.

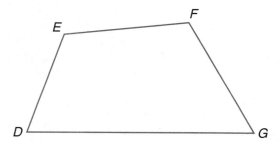

3. Construct a triangle that is the same shape as triangle *ABC* below but has sides twice as long.

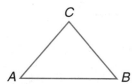

LESSON 5·8

Compass-and-Straightedge Constructions *cont.*

Use only a compass, a straightedge, and a sharp pencil. Use rulers and protractors only to check your work. Do not trace.

4. Construct a perpendicular bisector of this line segment.

_____

5. Divide this line segment into four equal parts.

_____

6. To **inscribe** a square in a circle means to construct a square inside a circle so that all four vertices (corners) of the square are on the circle. Draw a circle. Then inscribe a square in it.

7. Use your Geometry Template to draw a parallelogram. Then construct a line segment to show the height of the parallelogram. (That is, construct a perpendicular segment from one side to the opposite side.)

Angle Measures

Math Message

Write the measures of the angles indicated in Problems 1–6.
Do not use a protractor.

1. m∠a = _____

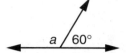

2. m∠x = _____   m∠y = _____   m∠z = _____

3. m∠p = _____

4. m∠r = _____   m∠s = _____   m∠t = _____

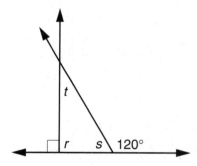

5. m∠h = _____

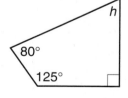

6. m∠d = _____   m∠e = _____

   m∠f = _____   m∠g = _____

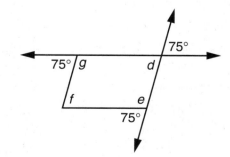

LESSON 5·9   Parallel Lines and Angle Relationships

Without using a protractor, find the degree measure of each angle in Problems 1–6 below. Write the measure inside the angle. Then circle the figures in which 2 of the lines appear to be parallel.

1.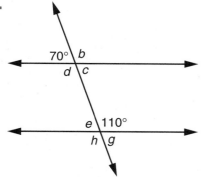

2.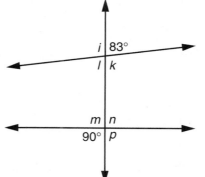

3.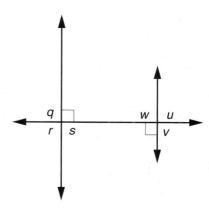

4.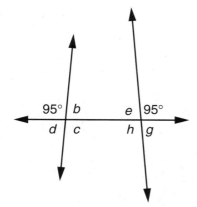

5.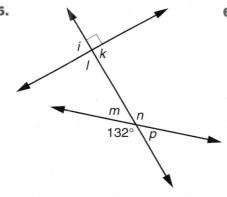

6.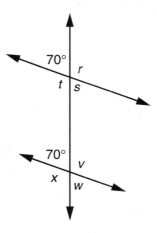

7. A line that intersects 2 parallel lines is called a **transversal.** The angles formed by 2 parallel lines and a transversal have special properties. Refer to the picture of parallel lines below to describe one of these properties.

Example: Angles such as ∠b and ∠f, which lie on the same side of the transversal, have the same measure.

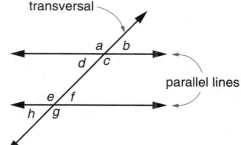

LESSON 5·9 Working with Parallel Lines

1. Using only a compass and a straightedge, construct 2 parallel lines. Do this construction without referring to the *Student Reference Book*. (*Hint:* This construction involves copying an angle.)

2. Draw 2 parallel lines using only a ruler and a pencil.

3. Draw a parallelogram that is not a rectangle, using only a ruler and a pencil.

LESSON 5·9 Calculating Total Price

A total price is the sum of the price of an item (or subtotal) and the sales tax or tip that is a percentage of that item:

Total Price = Subtotal + Sales Tax (or Tip)

Two-Step Method	One-Step Method
Subtotal: $49.75 Sales Tax: 8% Find the total price.	Subtotal: $49.75 Sales Tax: 8% Find the total price.
Step 1: Find the sales tax in dollars: 8% of $49.75. $0.08 * \$49.75 = \$3.98$	The total price equals 100% of the subtotal plus 8% of the subtotal, so
Step 2: Add the sales tax amount to the subtotal.	Total = 100% * subtotal + 8% * subtotal = 1.08 * subtotal
Subtotal + Sales Tax = Total Price $\$49.75 + \$3.98 = \$53.73$	Find 108% of $49.75. 108% of $49.75 = 1.08 * $49.75 = $53.73
The total price is $53.73.	The total price is $53.73.

Use either method to find the total price. Round your answer to the nearest cent, if necessary.

1. Subtotal: $89.00
   Sales Tax: 6%
   Total Price: _____

2. Subtotal: $325.00
   Sales Tax: 7%
   Total Price: _____

3. Subtotal: $25.20
   Tip: 15%
   Total Price: _____

4. Subtotal: $103.50
   Tip: 20%
   Total Price: _____

5. Subtotal: $448.40
   Sales Tax: 4.5%
   Total Price: _____

6. Subtotal: $876.00
   Sales Tax: 6.25%
   Total Price: _____

LESSON 5·9

Math Boxes

1. Reflect figure *PQST* over the *x*-axis. Then plot and label the vertices of the image that results from that reflection.

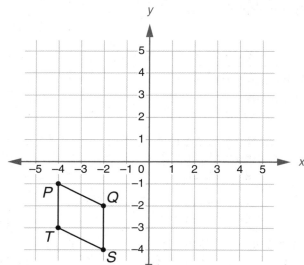

$P' = ($ _____ , _____ $)$ $Q' = ($ _____ , _____ $)$

$S' = ($ _____ , _____ $)$ $T' = ($ _____ , _____ $)$

SRB 180 234

2. Use the partial-quotients algorithm to divide the numerator by the denominator. Round the result to the nearest hundredth and rename the result as a percent.

$\frac{11}{12} = 0.$ _____ $=$ _____ %

SRB 55–57

3. Choose the best estimate for the product $11\frac{2}{3} * \frac{1}{4}$.

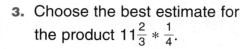 12

⬭ 11

⬭ 6

⬭ 3

SRB 90

4. Write a number sentence for each word sentence. Then tell whether the number sentence is true or false.

Word Sentence	Number Sentence	True or False?
If 19 is subtracted from 55, the result is 36.		
78 added to 62 is less than 160.		
45 is 5 times as great as 9.		

SRB 241–243

LESSON 5·10 Properties of Parallelograms

1. The 2 lines below are parallel. Place a ruler so that it intersects the parallel lines. Draw 2 transversals by drawing a line along each of the longer sides of the ruler. These lines should be parallel. The 4 lines form a parallelogram. Label the parallelogram *ABCD*.

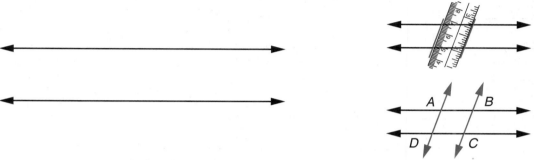

2. Use a compass to make sure that both pairs of opposite sides of your parallelogram are congruent.

3. Measure ∠*BAD* with a protractor. Write the measure on your drawing.

4. Find the measures of the other 3 angles of the parallelogram without using a protractor. Write the measures on your drawing.

5. Check your answers by measuring the angles with a protractor.

6. **a.** Angles *R* and *S* in the parallelogram at the right are opposite each other. Name another pair of opposite angles in the parallelogram.

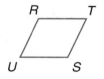

   _____

   b. What do you think is true about the opposite angles in a parallelogram?

   _____

7. **a.** Two angles of a polygon that are next to each other are called **consecutive angles.** Consecutive angles have a common side. Angles *R* and *U* in parallelogram *RUST* are consecutive angles. Name 3 other pairs of consecutive angles in this parallelogram.

   _____

   b. What do you think is true about consecutive angles in a parallelogram?

   _____

   _____

LESSON 5·10   # Parallelogram Problems

All figures on this page are parallelograms. Do not use a ruler or protractor to solve the problems below.

1. a. The measure of ∠A is _____. Explain how you know.

   _____

   _____

   _____

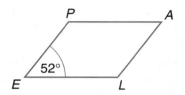

   b. The measure of ∠L is _____. Explain how you know.

   _____

   _____

   _____

2. The length of $\overline{AB}$ is _____. Explain how you know.

   _____

   _____

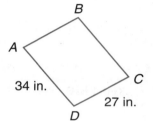

3. The measure of ∠OPT is _____. Explain how you know.

   _____

   _____

   _____

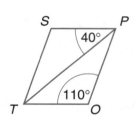

4. Quadrilateral *REST* is a square.

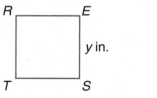

   a. Measure of ∠R = _____

   b. Length of $\overline{ES}$ = _____

   c. Perimeter of
      square *REST* = _____

5. What is the perimeter of parallelogram *BOYS*? *Hint:* First find the value of x.

   _____

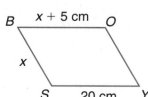

LESSON 5·10

Parallelogram Problems *continued*

6. Construct a parallelogram using only a compass and a straightedge. Do this construction without referring to page 191 of the *Student Reference Book*.

Try This

7. Quadrilaterals *FAIR* and *FARE* are parallelograms. Without using a protractor, find the measure of the angles in parallelogram *FAIR*.

What is the measure of

∠*IAF*? _____   ∠*AIR*? _____

∠*IRF*? _____   ∠*RFA*? _____

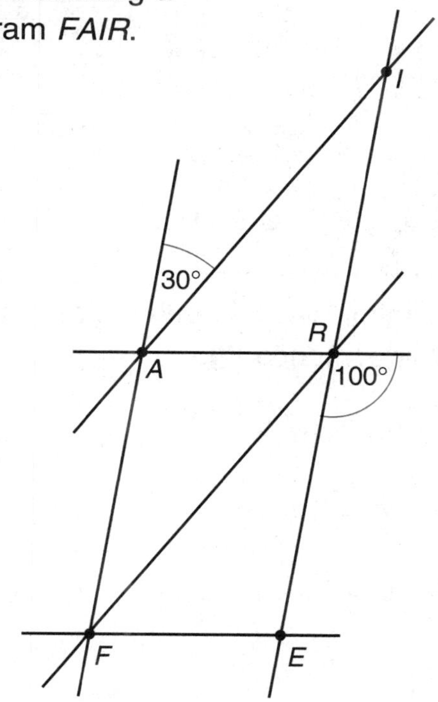

LESSON 5·10 Math Boxes

---

1. Without using a protractor, find the degree measure of each angle listed below.

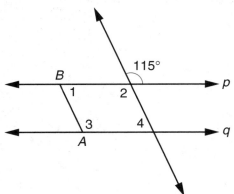

Lines *p* and *q* are parallel.

m∠1 = _____

m∠2 = _____

m∠3 = _____

m∠4 = _____

SRB 163

---

2. The advertising budget for a small company is shown in the table below. Complete the table. Then use a protractor to make a circle graph of the information. Title the graph.

Advertising Method	Amount of Money	Percent of Total	Degree Measure of Sector
Internet	$12,500		
Television		25%	90°
Radio	$18,750		
Magazines		45%	162°
Mailings	$6,250		
TOTAL	$125,000	100%	360°

_____

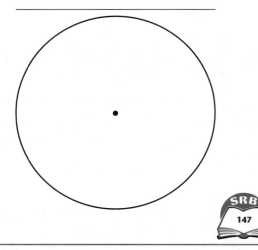

SRB 147

---

3. Find the missing value that makes the number sentence true.

a. $\frac{1}{5} * m = 10$ m = _____

b. $p \div 4 = 1.5$ p = _____

c. $5\frac{1}{5} * 1\frac{1}{2} = k$ k = _____

SRB 87–90

4. How many $\frac{3}{16}$-inch segments are in a line segment that is $1\frac{1}{8}$-inches long? Fill in the circle next to the best answer.

Ⓐ 6   Ⓑ $7\frac{1}{2}$

Ⓒ 9   Ⓓ 18

SRB 91

LESSON 5·11 **Math Boxes**

1. Find the missing value that makes the number sentence true.

   a. $18 + n = 37$ $n =$ _____

   b. $5 * (x - 2) = 10$ $x =$ _____

   c. $6 = \left(\dfrac{b}{3}\right) + 6$ $b =$ _____

   d. $t - 4.5 = 5.5$ $t =$ _____

242 243

2. Multiply or divide.

   a. _____ $= 80 * \dfrac{1}{4}$

   b. $75 \div 5 =$ _____

   c. $\dfrac{3}{5} * 50 =$ _____

   d. _____ $= (5 * 60) \div 6$

88

3. Make each sentence true by inserting parentheses.

   a. $5 + 5 - 3 * \dfrac{6}{6} = 7$

   b. $3 * 9 - 5 + \dfrac{8}{2} = 26$

   c. $36 / 6 + 3 - 3^2 = 0$

   d. $1\dfrac{3}{4} - \dfrac{1}{2} + \dfrac{5}{8} = 1\dfrac{7}{8}$

   e. $3\dfrac{1}{2} - 1 + 1\dfrac{1}{4} = \dfrac{15}{4}$

247

4. Multiply. Write your answer in simplest form.

   a. _____ $= \dfrac{3}{5} * \dfrac{2}{8}$

   b. _____ $= \dfrac{4}{9} * \dfrac{3}{6}$

   c. $4 * 6\dfrac{3}{10} =$ _____

   d. $2\dfrac{2}{3} * 5\dfrac{1}{8} =$ _____

   e. _____ $= \dfrac{9}{11} * \dfrac{5}{6}$

88–90

5. Draw a line segment that is 3 inches long.

How many $\dfrac{1}{2}$-inch segments are in 3 inches?

_____

91

6. Compare using $<$, $>$, or $=$.

   a. $4^2 + (6 * 7)$ _____ $(4^2 + 6) * 7$

   b. $4.5 \div 3$ _____ $3 \div 4.5$

   c. $48 / (12 - 6)$ _____ $(48 / 12) - 6$

   d. $(25 * 3) \div 5$ _____ $25 * \dfrac{3}{5}$

104

Reference

Metric System

Units of Length
1 kilometer (km)	= 1,000 meters (m)
1 meter	= 10 decimeters (dm)
	= 100 centimeters (cm)
	= 1,000 millimeters (mm)
1 decimeter	= 10 centimeters
1 centimeter	= 10 millimeters

Units of Area
1 square meter (m^2)	= 100 square decimeters (dm^2)
	= 10,000 square centimeters (cm^2)
1 square decimeter	= 100 square centimeters
1 are (a)	= 100 square meters
1 hectare (ha)	= 100 ares
1 square kilometer (km^2)	= 100 hectares

Units of Volume
1 cubic meter (m^3)	= 1,000 cubic decimeters (dm^3)
	= 1,000,000 cubic centimeters (cm^3)
1 cubic decimeter	= 1,000 cubic centimeters

Units of Capacity
1 kiloliter (kL)	= 1,000 liters (L)
1 liter	= 1,000 milliliters (mL)

Units of Mass
1 metric ton (t)	= 1,000 kilograms (kg)
1 kilogram	= 1,000 grams (g)
1 gram	= 1,000 milligrams (mg)

Units of Time
1 century	= 100 years
1 decade	= 10 years
1 year (yr)	= 12 months
	= 52 weeks (plus one or two days)
	= 365 days (366 days in a leap year)
1 month (mo)	= 28, 29, 30, or 31 days
1 week (wk)	= 7 days
1 day (d)	= 24 hours
1 hour (hr)	= 60 minutes
1 minute (min)	= 60 seconds (sec)

U.S. Customary System

Units of Length
1 mile (mi)	= 1,760 yards (yd)
	= 5,280 feet (ft)
1 yard	= 3 feet
	= 36 inches (in.)
1 foot	= 12 inches

Units of Area
1 square yard (yd^2)	= 9 square feet (ft^2)
	= 1,296 square inches ($in.^2$)
1 square foot	= 144 square inches
1 acre	= 43,560 square feet
1 square mile (mi^2)	= 640 acres

Units of Volume
1 cubic yard (yd^3)	= 27 cubic feet (ft^3)
1 cubic foot	= 1,728 cubic inches ($in.^3$)

Units of Capacity
1 gallon (gal)	= 4 quarts (qt)
1 quart	= 2 pints (pt)
1 pint	= 2 cups (c)
1 cup	= 8 fluid ounces (fl oz)
1 fluid ounce	= 2 tablespoons (tbs)
1 tablespoon	= 3 teaspoons (tsp)

Units of Weight
1 ton (T)	= 2,000 pounds (lb)
1 pound	= 16 ounces (oz)

System Equivalents

1 inch is about 2.5 cm (2.54).

1 kilometer is about 0.6 mile (0.621).

1 mile is about 1.6 kilometers (1.609).

1 meter is about 39 inches (39.37).

1 liter is about 1.1 quarts (1.057).

1 ounce is about 28 grams (28.350).

1 kilogram is about 2.2 pounds (2.205).

1 hectare is about 2.5 acres (2.47).

Rules for Order of Operations

1. Do operations within parentheses or other grouping symbols before doing anything else.
2. Calculate all exponents.
3. Multiply or divide in order from left to right.
4. Add or subtract in order from left to right.

Reference

Symbols

+	plus or positive
−	minus or negative
*, ×	multiplied by
÷, /	divided by
=	is equal to
≠	is not equal to
<	is less than
>	is greater than
≤	is less than or equal to
≥	is greater than or equal to
x^n	nth power of x
$\sqrt{x}$	square root of x
%	percent
a:b, a / b, $\frac{a}{b}$	ratio of a to b or a divided by b or the fraction $\frac{a}{b}$
°	degree
(a,b)	ordered pair
$\overleftrightarrow{AS}$	line AS
$\overline{AS}$	line segment AS
$\overrightarrow{AS}$	ray AS
∟	right angle
⊥	is perpendicular to
∥	is parallel to
△ABC	triangle ABC
∠ABC	angle ABC
∠B	angle B

Place-Value Chart

trillions	100B	10B	billions	100M	10M	millions	hundred-thousands	ten-thousands	thousands	hundreds	tens	ones	.	tenths	hundredths	thousandths
1,000 billions			1,000 millions			1,000,000s	100,000s	10,000s	1,000s	100s	10s	1s	.	0.1s	0.01s	0.001s
10^{12}	10^{11}	10^{10}	10^{9}	10^{8}	10^{7}	10^{6}	10^{5}	10^{4}	10^{3}	10^{2}	10^{1}	10^{0}	.	10^{-1}	10^{-2}	10^{-3}

Probability Meter

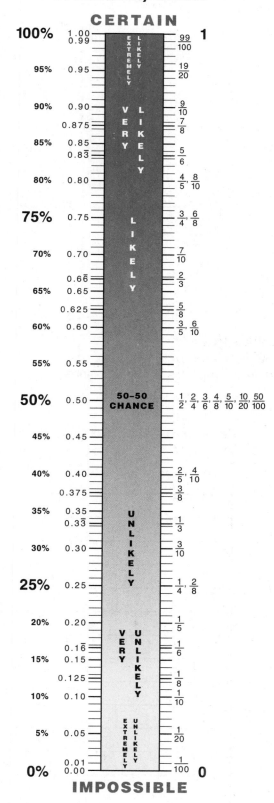

Reference

Latitude and Longitude

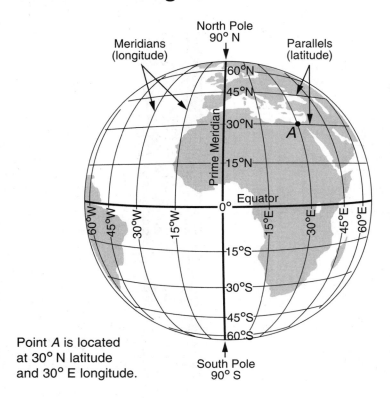

Point *A* is located at 30° N latitude and 30° E longitude.

Rational Numbers

Rule	Example
$\dfrac{a}{b} = \dfrac{n*a}{n*b}$	$\dfrac{2}{3} = \dfrac{4*2}{4*3} = \dfrac{8}{12}$
$\dfrac{a}{b} = \dfrac{a/n}{b/n}$	$\dfrac{8}{12} = \dfrac{8/4}{12/4} = \dfrac{2}{3}$
$\dfrac{a}{a} = a * \dfrac{1}{a} = 1$	$\dfrac{4}{4} = 4 * \dfrac{1}{4} = 1$
$\dfrac{a}{b} + \dfrac{c}{b} = \dfrac{a+c}{b}$	$\dfrac{3}{5} + \dfrac{1}{5} = \dfrac{3+1}{5} = \dfrac{4}{5}$
$\dfrac{a}{b} - \dfrac{c}{b} = \dfrac{a-c}{b}$	$\dfrac{3}{5} - \dfrac{1}{5} = \dfrac{3-1}{5} = \dfrac{2}{5}$
$\dfrac{a}{b} * \dfrac{c}{d} = \dfrac{a*c}{b*d}$	$\dfrac{1}{4} * \dfrac{2}{3} = \dfrac{1*2}{4*3} = \dfrac{2}{12}$

To compare, add, or subtract fractions:
1. Find a common denominator.
2. Rewrite fractions as equivalent fractions with the common denominator.
3. Compare, add, or subtract these fractions.

Fraction-Stick and Decimal Number-Line Chart

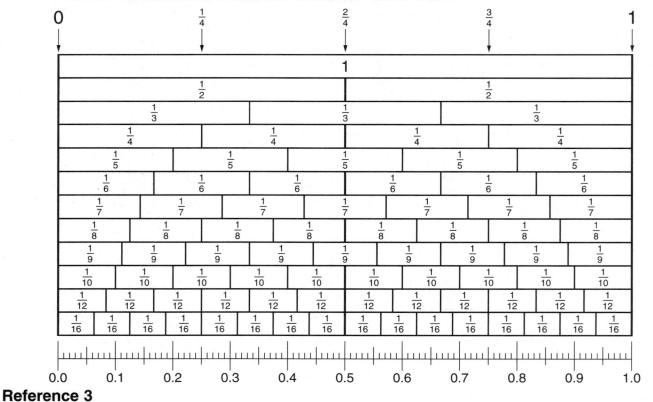

Reference

Equivalent Fractions, Decimals, and Percents

															Decimal	Percent
$\frac{1}{2}$	$\frac{2}{4}$	$\frac{3}{6}$	$\frac{4}{8}$	$\frac{5}{10}$	$\frac{6}{12}$	$\frac{7}{14}$	$\frac{8}{16}$	$\frac{9}{18}$	$\frac{10}{20}$	$\frac{11}{22}$	$\frac{12}{24}$	$\frac{13}{26}$	$\frac{14}{28}$	$\frac{15}{30}$	0.5	50%
$\frac{1}{3}$	$\frac{2}{6}$	$\frac{3}{9}$	$\frac{4}{12}$	$\frac{5}{15}$	$\frac{6}{18}$	$\frac{7}{21}$	$\frac{8}{24}$	$\frac{9}{27}$	$\frac{10}{30}$	$\frac{11}{33}$	$\frac{12}{36}$	$\frac{13}{39}$	$\frac{14}{42}$	$\frac{15}{45}$	$0.\overline{3}$	$33\frac{1}{3}\%$
$\frac{2}{3}$	$\frac{4}{6}$	$\frac{6}{9}$	$\frac{8}{12}$	$\frac{10}{15}$	$\frac{12}{18}$	$\frac{14}{21}$	$\frac{16}{24}$	$\frac{18}{27}$	$\frac{20}{30}$	$\frac{22}{33}$	$\frac{24}{36}$	$\frac{26}{39}$	$\frac{28}{42}$	$\frac{30}{45}$	$0.\overline{6}$	$66\frac{2}{3}\%$
$\frac{1}{4}$	$\frac{2}{8}$	$\frac{3}{12}$	$\frac{4}{16}$	$\frac{5}{20}$	$\frac{6}{24}$	$\frac{7}{28}$	$\frac{8}{32}$	$\frac{9}{36}$	$\frac{10}{40}$	$\frac{11}{44}$	$\frac{12}{48}$	$\frac{13}{52}$	$\frac{14}{56}$	$\frac{15}{60}$	0.25	25%
$\frac{3}{4}$	$\frac{6}{8}$	$\frac{9}{12}$	$\frac{12}{16}$	$\frac{15}{20}$	$\frac{18}{24}$	$\frac{21}{28}$	$\frac{24}{32}$	$\frac{27}{36}$	$\frac{30}{40}$	$\frac{33}{44}$	$\frac{36}{48}$	$\frac{39}{52}$	$\frac{42}{56}$	$\frac{45}{60}$	0.75	75%
$\frac{1}{5}$	$\frac{2}{10}$	$\frac{3}{15}$	$\frac{4}{20}$	$\frac{5}{25}$	$\frac{6}{30}$	$\frac{7}{35}$	$\frac{8}{40}$	$\frac{9}{45}$	$\frac{10}{50}$	$\frac{11}{55}$	$\frac{12}{60}$	$\frac{13}{65}$	$\frac{14}{70}$	$\frac{15}{75}$	0.2	20%
$\frac{2}{5}$	$\frac{4}{10}$	$\frac{6}{15}$	$\frac{8}{20}$	$\frac{10}{25}$	$\frac{12}{30}$	$\frac{14}{35}$	$\frac{16}{40}$	$\frac{18}{45}$	$\frac{20}{50}$	$\frac{22}{55}$	$\frac{24}{60}$	$\frac{26}{65}$	$\frac{28}{70}$	$\frac{30}{75}$	0.4	40%
$\frac{3}{5}$	$\frac{6}{10}$	$\frac{9}{15}$	$\frac{12}{20}$	$\frac{15}{25}$	$\frac{18}{30}$	$\frac{21}{35}$	$\frac{24}{40}$	$\frac{27}{45}$	$\frac{30}{50}$	$\frac{33}{55}$	$\frac{36}{60}$	$\frac{39}{65}$	$\frac{42}{70}$	$\frac{45}{75}$	0.6	60%
$\frac{4}{5}$	$\frac{8}{10}$	$\frac{12}{15}$	$\frac{16}{20}$	$\frac{20}{25}$	$\frac{24}{30}$	$\frac{28}{35}$	$\frac{32}{40}$	$\frac{36}{45}$	$\frac{40}{50}$	$\frac{44}{55}$	$\frac{48}{60}$	$\frac{52}{65}$	$\frac{56}{70}$	$\frac{60}{75}$	0.8	80%
$\frac{1}{6}$	$\frac{2}{12}$	$\frac{3}{18}$	$\frac{4}{24}$	$\frac{5}{30}$	$\frac{6}{36}$	$\frac{7}{42}$	$\frac{8}{48}$	$\frac{9}{54}$	$\frac{10}{60}$	$\frac{11}{66}$	$\frac{12}{72}$	$\frac{13}{78}$	$\frac{14}{84}$	$\frac{15}{90}$	$0.1\overline{6}$	$16\frac{2}{3}\%$
$\frac{5}{6}$	$\frac{10}{12}$	$\frac{15}{18}$	$\frac{20}{24}$	$\frac{25}{30}$	$\frac{30}{36}$	$\frac{35}{42}$	$\frac{40}{48}$	$\frac{45}{54}$	$\frac{50}{60}$	$\frac{55}{66}$	$\frac{60}{72}$	$\frac{65}{78}$	$\frac{70}{84}$	$\frac{75}{90}$	$0.8\overline{3}$	$83\frac{1}{3}\%$
$\frac{1}{7}$	$\frac{2}{14}$	$\frac{3}{21}$	$\frac{4}{28}$	$\frac{5}{35}$	$\frac{6}{42}$	$\frac{7}{49}$	$\frac{8}{56}$	$\frac{9}{63}$	$\frac{10}{70}$	$\frac{11}{77}$	$\frac{12}{84}$	$\frac{13}{91}$	$\frac{14}{98}$	$\frac{15}{105}$	0.143	14.3%
$\frac{2}{7}$	$\frac{4}{14}$	$\frac{6}{21}$	$\frac{8}{28}$	$\frac{10}{35}$	$\frac{12}{42}$	$\frac{14}{49}$	$\frac{16}{56}$	$\frac{18}{63}$	$\frac{20}{70}$	$\frac{22}{77}$	$\frac{24}{84}$	$\frac{26}{91}$	$\frac{28}{98}$	$\frac{30}{105}$	0.286	28.6%
$\frac{3}{7}$	$\frac{6}{14}$	$\frac{9}{21}$	$\frac{12}{28}$	$\frac{15}{35}$	$\frac{18}{42}$	$\frac{21}{49}$	$\frac{24}{56}$	$\frac{27}{63}$	$\frac{30}{70}$	$\frac{33}{77}$	$\frac{36}{84}$	$\frac{39}{91}$	$\frac{42}{98}$	$\frac{45}{105}$	0.429	42.9%
$\frac{4}{7}$	$\frac{8}{14}$	$\frac{12}{21}$	$\frac{16}{28}$	$\frac{20}{35}$	$\frac{24}{42}$	$\frac{28}{49}$	$\frac{32}{56}$	$\frac{36}{63}$	$\frac{40}{70}$	$\frac{44}{77}$	$\frac{48}{84}$	$\frac{52}{91}$	$\frac{56}{98}$	$\frac{60}{105}$	0.571	57.1%
$\frac{5}{7}$	$\frac{10}{14}$	$\frac{15}{21}$	$\frac{20}{28}$	$\frac{25}{35}$	$\frac{30}{42}$	$\frac{35}{49}$	$\frac{40}{56}$	$\frac{45}{63}$	$\frac{50}{70}$	$\frac{55}{77}$	$\frac{60}{84}$	$\frac{65}{91}$	$\frac{70}{98}$	$\frac{75}{105}$	0.714	71.4%
$\frac{6}{7}$	$\frac{12}{14}$	$\frac{18}{21}$	$\frac{24}{28}$	$\frac{30}{35}$	$\frac{36}{42}$	$\frac{42}{49}$	$\frac{48}{56}$	$\frac{54}{63}$	$\frac{60}{70}$	$\frac{66}{77}$	$\frac{72}{84}$	$\frac{78}{91}$	$\frac{84}{98}$	$\frac{90}{105}$	0.857	85.7%
$\frac{1}{8}$	$\frac{2}{16}$	$\frac{3}{24}$	$\frac{4}{32}$	$\frac{5}{40}$	$\frac{6}{48}$	$\frac{7}{56}$	$\frac{8}{64}$	$\frac{9}{72}$	$\frac{10}{80}$	$\frac{11}{88}$	$\frac{12}{96}$	$\frac{13}{104}$	$\frac{14}{112}$	$\frac{15}{120}$	0.125	$12\frac{1}{2}\%$
$\frac{3}{8}$	$\frac{6}{16}$	$\frac{9}{24}$	$\frac{12}{32}$	$\frac{15}{40}$	$\frac{18}{48}$	$\frac{21}{56}$	$\frac{24}{64}$	$\frac{27}{72}$	$\frac{30}{80}$	$\frac{33}{88}$	$\frac{36}{96}$	$\frac{39}{104}$	$\frac{42}{112}$	$\frac{45}{120}$	0.375	$37\frac{1}{2}\%$
$\frac{5}{8}$	$\frac{10}{16}$	$\frac{15}{24}$	$\frac{20}{32}$	$\frac{25}{40}$	$\frac{30}{48}$	$\frac{35}{56}$	$\frac{40}{64}$	$\frac{45}{72}$	$\frac{50}{80}$	$\frac{55}{88}$	$\frac{60}{96}$	$\frac{65}{104}$	$\frac{70}{112}$	$\frac{75}{120}$	0.625	$62\frac{1}{2}\%$
$\frac{7}{8}$	$\frac{14}{16}$	$\frac{21}{24}$	$\frac{28}{32}$	$\frac{35}{40}$	$\frac{42}{48}$	$\frac{49}{56}$	$\frac{56}{64}$	$\frac{63}{72}$	$\frac{70}{80}$	$\frac{77}{88}$	$\frac{84}{96}$	$\frac{91}{104}$	$\frac{98}{112}$	$\frac{105}{120}$	0.875	$87\frac{1}{2}\%$
$\frac{1}{9}$	$\frac{2}{18}$	$\frac{3}{27}$	$\frac{4}{36}$	$\frac{5}{45}$	$\frac{6}{54}$	$\frac{7}{63}$	$\frac{8}{72}$	$\frac{9}{81}$	$\frac{10}{90}$	$\frac{11}{99}$	$\frac{12}{108}$	$\frac{13}{117}$	$\frac{14}{126}$	$\frac{15}{135}$	$0.\overline{1}$	$11\frac{1}{9}\%$
$\frac{2}{9}$	$\frac{4}{18}$	$\frac{6}{27}$	$\frac{8}{36}$	$\frac{10}{45}$	$\frac{12}{54}$	$\frac{14}{63}$	$\frac{16}{72}$	$\frac{18}{81}$	$\frac{20}{90}$	$\frac{22}{99}$	$\frac{24}{108}$	$\frac{26}{117}$	$\frac{28}{126}$	$\frac{30}{135}$	$0.\overline{2}$	$22\frac{2}{9}\%$
$\frac{4}{9}$	$\frac{8}{18}$	$\frac{12}{27}$	$\frac{16}{36}$	$\frac{20}{45}$	$\frac{24}{54}$	$\frac{28}{63}$	$\frac{32}{72}$	$\frac{36}{81}$	$\frac{40}{90}$	$\frac{44}{99}$	$\frac{48}{108}$	$\frac{52}{117}$	$\frac{56}{126}$	$\frac{60}{135}$	$0.\overline{4}$	$44\frac{4}{9}\%$
$\frac{5}{9}$	$\frac{10}{18}$	$\frac{15}{27}$	$\frac{20}{36}$	$\frac{25}{45}$	$\frac{30}{54}$	$\frac{35}{63}$	$\frac{40}{72}$	$\frac{45}{81}$	$\frac{50}{90}$	$\frac{55}{99}$	$\frac{60}{108}$	$\frac{65}{117}$	$\frac{70}{126}$	$\frac{75}{135}$	$0.\overline{5}$	$55\frac{5}{9}\%$
$\frac{7}{9}$	$\frac{14}{18}$	$\frac{21}{27}$	$\frac{28}{36}$	$\frac{35}{45}$	$\frac{42}{54}$	$\frac{49}{63}$	$\frac{56}{72}$	$\frac{63}{81}$	$\frac{70}{90}$	$\frac{77}{99}$	$\frac{84}{108}$	$\frac{91}{117}$	$\frac{98}{126}$	$\frac{105}{135}$	$0.\overline{7}$	$77\frac{7}{9}\%$
$\frac{8}{9}$	$\frac{16}{18}$	$\frac{24}{27}$	$\frac{32}{36}$	$\frac{40}{45}$	$\frac{48}{54}$	$\frac{56}{63}$	$\frac{64}{72}$	$\frac{72}{81}$	$\frac{80}{90}$	$\frac{88}{99}$	$\frac{96}{108}$	$\frac{104}{117}$	$\frac{112}{126}$	$\frac{120}{135}$	$0.\overline{8}$	$88\frac{8}{9}\%$

Note: The decimals for sevenths have been rounded to the nearest thousandth.

Date

Time

Spoon Scramble Cards 1

$\dfrac{1}{7}$ of 42	$\dfrac{24}{4} * \dfrac{5}{5}$	$\dfrac{54}{9}$	$2\dfrac{16}{4}$
$\dfrac{1}{5}$ of 35	$\dfrac{21}{3} * \dfrac{4}{4}$	$\dfrac{56}{8}$	$4\dfrac{36}{12}$
$\dfrac{1}{8}$ of 64	$\dfrac{48}{6} * \dfrac{3}{3}$	$\dfrac{32}{4}$	$3\dfrac{25}{5}$
$\dfrac{1}{4}$ of 36	$\dfrac{63}{7} * \dfrac{6}{6}$	$\dfrac{72}{8}$	$5\dfrac{32}{8}$

Activity Sheet 1

Spoon Scramble Cards 2

$1 \div 2$	$\dfrac{35}{70}$	$\dfrac{1}{8} * 4$	0.5
$\dfrac{1}{3}$	$\dfrac{1}{6} * 2$	$33\dfrac{1}{3}\%$	$\dfrac{1}{2} - \dfrac{1}{6}$
$\dfrac{26}{13}$	$\left(\dfrac{6}{9} * \dfrac{9}{6}\right) * 2$	2	$4 * \dfrac{1}{2}$
$\dfrac{3}{4}$	$\dfrac{600}{800}$	0.75	$3 \div 4$

Algebra Election Cards, Set 1

Find:

x squared

x to the fourth power

$\dfrac{1}{x}$

Find n. (*Hint:* n could be a negative number.)

$1{,}000 + n = x$

$1{,}000 + n = -x$

Complete.

$x * 10^6 = $ _____ million

$x * 10^9 = $ _____ billion

$x * 10^{12} = $ _____ _____

What is the value of n?

$-20 + x = n$

$-100 + (-x) = n$

Insert parentheses in

$\dfrac{1}{10} * x - 2$

so that its value is greater than 0 and less than 4.

Find n. (*Hint:* n could be a negative number.)

$n + 10 = x$

$n - 10 = x$

What is the value of n?

$n = ((5 * x) - 4) / 2$

What is the value of n?

$20 + (-x) = n$

$-20 - (-x) = n$

If $B = 80$ and $H = 100x$, what does T equal?

$T = B - (2 * \dfrac{H}{1{,}000})$

Find n.

$n = (2 * x) / 10$

$n + 1 = (2 * x)$

Suppose you earn x dollars per hour. Complete the table.

Time	Earnings
1 hr	$
2 hr	$
4 hr	$
10 hr	$

Which is greater:

x^2 or 10^3?

x^3 or 10^4?

Tell whether each is true or false.

$10 * x > 100$

$\dfrac{1}{2} * x * 100 < 10^3$

$x^3 * 1{,}000 > 4 * 10^4$

Which number is this?

$x * 10^2$

$x * 10^5$

A boulder dropped off a cliff falls approximately $16 * x^2$ feet in x seconds. How many feet is that?

Which is less:

$\dfrac{x^3}{10}$ or $(x + 10)^2$?

$10 * x^2$ or $(x + 10)^3$?

Activity Sheet 3

Algebra Election Cards, Set 2

What is n? $5 + 2 * x = n + x$	Tell which is correct for each: $<$, $=$, or $>$. $x < = > 30 - x$ $x < = > 20 - x$ $x < = > 10 - x$
$x + \triangle$ 200 oz 1 $\triangle$ weighs _____ ounces.	Name a number n such that $x - n$ is a negative number greater than -10.
Find w. $w = x^2$ $w = x * 10^0$	Suppose you have 10 $\boxed{+}$ markers and $2 * x$ $\boxed{-}$ markers. What is your balance?
Is point (x, y) above, below, or on the line through points A and B? $A (0,30)$ $B (60,30)$	Suppose you have x $\boxed{+}$ markers and 40 $\boxed{-}$ markers. What is your balance?

Is point (x, y) to the left of, to the right of, or on the line through points A and B? $B (30,60)$ $A (30,0)$	Is $\frac{1}{x}$ greater than, less than, or equal to $\frac{1}{10}$?
What is the value of n? $10 + (-x) = n$ $-10 - (-x) = n$	Subtract. $x - 100 = ?$ $x - (-100) = ?$
What is the median of 4, 8, 12, 13, and x?	Add. $-25 + x = ?$ $x + 3 - 10 = ?$
If $(2 * x) + n = 100$, what is the value of n?	Suppose you travel x miles per hour. Complete the table. | Time	Distance
1 hr	
2 hr	
4 hr	
10 hr		

Activity Sheet 4